THE SILVER LINK LIBRARY OF
RAILWAY MODELLING

●

BASEBOARD BASICS
and
MAKING TRACK

THE SILVER LINK LIBRARY OF RAILWAY MODELLING

BASEBOARD BASICS
and
MAKING TRACKS

Planning * Baseboard construction * Track-laying * Wiring

A series of books providing a no-nonsense step-by-step guide to model railway construction

Trevor Booth

Silver Link Publishing Ltd

First published in 1993
Reprinted 1995
Reprinted 1997
Reprinted 1999
Reprinted 2000
Reprinted 2001
Reprinted 2002
Reprinted 2004
Reprinted 2004
Reprinted 2008

Electricity can be dangerous! Great care must always be taken when handling or assembling electrical equipment, and neither the publisher nor the author can accept responsibility for any accidents that may occur. If in doubt, always consult a qualified electrician.

British Library Cataloguing in Publication Data

A catalogue record for this book is available from the British Library.

ISBN 978 1 85794 006 0

Silver Link Publishing Ltd
The Trundle
Ringstead Road
Great Addington
Kettering
Northants
NN14 4BW

Tel/Fax: 01536 330588
email: sales@nostalgiacollection.com
Website: www.nostalgiacollection.com

Printed and bound in the Czech Republic

ACKNOWLEDGEMENTS

Grateful thanks are due to Peter Rogers, Peter Smith, David Hampson, John Harmon and the staff of Salford Quays Heritage Centre and Bolton Local History Library, who have helped practically with photographs and information.

During several years' interest in railway modelling there have been many, many people who have willingly offered help, guidance and information, and who have passed on their knowledge and skills - their help and encouragement is acknowledged and gratefully appreciated.

Special thanks go to my wife Susan for typing the manuscript and making sense of my scribbled notes, providing moral support and encouragement, and tolerating the eccentricities of a railway modeller.

CONTENTS

INTRODUCTION

This is the first in a series of books dealing in detail with the construction of a model railway, covering in depth the totality of building from concept and design right the way through, step by step, to operating the completed model.

Unlike other books on the subject, this series looks at the construction of a specific model railway in detail. The process of construction is pursued via a number of key reference points, and while the core of the book relates to a specific model, alternatives and different approaches are discussed.

The layout chosen as the key reference point and built for this series is actually in 7 mm scale, O gauge. This both reflects the increasing popularity of what is surely the king of all modelling scales, and the fact that many of the techniques, principles and materials used are equally relevant to other scales, in particular 3.5/4 mm scale derivatives of HO, OO, EM, S4 and the narrow gauge variants. Indeed, having evolved my own modelling interests through OO gauge, EM, a brief dalliance with N scale and latterly O gauge, I unashamedly admit to using the techniques of the smaller scales in the larger!

In order that any layout project described has some relevance to the practical limitations of space, time and money available, the main layout and all the alternatives discussed and described are what I would suggest could be accommodated within these constraints for the 'average' model railroader.

The hobby of railway modelling has many byways, and specific interests and areas of expertise develop. I will not offer anything new or revolutionary in the techniques and materials outlined, but to use rather the tried and tested methods which I have found successful.

I hope that these books encourage and support, and that above all they provide enjoyment. If they help and assist as problems arise in the pursuit of the hobby, as they surely will, then I will be only too pleased, as they will then have achieved their purpose.

1.
WHAT WE HOPE TO ACHIEVE

The purpose of this book is to help the reader through the process of building a model railway. You won't find much in the way of discussion on the various scales or gauges, types of layout, etc, but hopefully you will find much practical detail on layout design, baseboard construction, track-laying and electrics, and in subsequent volumes scenic construction, buildings, structures and ultimately locomotives and rolling-stock.

In order to develop and demonstrate 'how to', the book, as a base, describes in detail the evolution of design and construction of a model railway. The particular model discussed is constructed in O gauge, 7 mm scale, but the techniques and materials used are fairly common to all the popular modelling scales. Where there are of necessity radically different techniques or materials needed in a particular scale, these are discussed. The aim is to show practical methods with the widest application.

The layout built for this project, which we have called Platt Lane, is an interesting exercise in the art of compromise, an essential concept in building a model railway. It represents the terminus of a railway running to a large northern town, and incorporates goods facilities, the potential for intensive operation and the use of a reasonable amount of stock and locomotives in a comparatively small area. There is also considerable potential for scenic development and, for those who like that sort of thing, potential for extensive signalling, interlocking and automatic operation.

The primary period for this demonstration project is the late 1950s/early 1960s. However, it can very easily be taken back in time or even brought forward to the diesel/electric era. Needless to say, the layout design can be adapted to represent virtually anywhere in the UK or even further afield and this will also be discussed. Similarly, the layout design and the size and shape can easily be adapted - it is really only a demonstration project.

'If it looks right, it is right'

Beginning at the beginning, 'layout design' is really a pretentious phrase for describing what you want from your model railway and working out how you are going to do it. It does, however, describes accurately the process of taking on an idea for a model railway, developing it through to something that will fit your requirements. It can be extended to cover the locomotives and rolling-stock to be used and the subsequent operation. The proposed use of the layout is also important.

I think, based on my own experiences, that there are two factors that need to be borne in mind in building a model railway: first, any model railway is a compromise, and second, the adage that 'if it looks right, it is right' is worth a lot more than slide rules and micrometers if you are going to achieve something that is visually 'correct'.

So far as the first factor is concerned, let me explain clearly what I mean. The last twenty years or so have seen an awful lot of articles in the model press on 'fine' or 'exact' scale modelling and the development of Scalefour, Scaleseven and 2 mm scale layouts. While the attitude of some of the proponents of these standards has been somewhat 'holier than thou', the spin-off has been the development of products of improved quality available to the railway modeller, and the fairly general availability of high-grade components. Perhaps the most significant effect has been the spread of this improvement to the standards of model railways being produced across *all* the scales and gauges, and particularly the spread of this desire for perfection as regards the scenery, structures and the presentation of the model railway, to match the refinements in the 'engineering' side of things.

However, despite greater dimensional accuracy of models, better presentation of the layout and exquisite representations of buildings, trees and herbage, the model railway is and can only ever be an illustration of a particular scene. No amount of modelling accuracy will ever overcome the fact that the locomotives are not in steam or running on diesel fuel, that the weight is not scaled down, that people and road vehicles do not move, to mention just a few anomalies and inevitable 'inaccuracies'. The compromise is always there. Nonetheless, the fine scale ethic has, I believe, brought to railway modelling a much greater awareness of the reality that our models are supposed to represent, and this has contributed greatly to the visual appearance of model railways.

This leads us nicely into the second factor I referred to, which I have from my own experience come to regard as perhaps the most important and which for want of a better description I would call the 'appearance factor'. No matter how accurate a model may be dimensionally, if it doesn't look the part then it simply won't do. I'm not always sure why this should be, but I recall building a Stanier 4,000-gallon tender in 4 mm scale some years ago, the principal aim of which was to achieve as accurate a model as possible. Much research and metal-bashing later, a tender was produced, but it didn't look quite right. It checked out dimensionally, but alas just did not look the part. So while keeping the basic length, width and height, I made slight alterations to the length of the raised coal-hole sides and altered their curvature judiciously until it looked right. These sides no longer matched the prototype profile exactly, but the tender looked the part.

I don't know why this should be, but it is also a phenomenon, if that's not too much of an exaggeration, that I have noticed particularly with buildings. Here, I have often found it necessary to juggle with scale to get structures to look right when in place on the layout. Whether this is due to the human eye and the unnatural angle from which models are viewed, I cannot say.

Where does this lead? Well, in my view, it is the overall picture we get from a model railway, the impression that it gives us, that is more important than the exact dimensional accuracy of each component; in effect, achieving an overall standard from trackwork via locomotives and stock to the scenery and structures that is consistent and evokes the character, flavour and general impression of the place and time being modelled. There is little point in having museum-quality models of locomotives pulling printed-card wagons past cornflake-packet station buildings - the illusion is lost (unless, of course, you have a tremendously strange imagination).

I often wonder whether this sense of consistency was one of the factors in the success and popularity of the old Hornby-Dublo range. There was a consistency in the quality across the range of locos, stock and buildings offered both in the tinplate era and later in the plastic super-detail period. They certainly weren't the most accurate of models, but for me they captured the flavour of the real thing better than many of the much more accurate models of their successors. Like many thousands of others, oh how I wish I had kept my childhood toys!

So you don't want a super-detail, accurate model railway, just somewhere to run and display your collection of models? Well, I would submit that they will look even better in surroundings which give the impression of a real railway! Even if you don't want to model a particular ex-Midland, Great Western or whatever branch line, there are some general points that can be followed which mirror the norm for railways in a particular area or country.

I think Cyril Freezer, former editor of *Railway Modeller*, suggested in an editorial some years ago that a model railway should be capable of telling its viewer its location and period without locomotives or rolling-stock, the obvious 'give-aways' being present. I think this is a very laudable aim. The visual juxtaposition of railway buildings, fads for a certain style of operation leading to a peculiar track layout, for example the Midland Railway's aversion to facing crossovers on main lines, the style of buildings or signals on a particular line - all these can easily achieve this result.

The source information for many of the world's railways is there, either still to be seen in the current railway scene or well documented in the increasing numbers of specialist magazines, books and, more recently, videos.

I firmly believe that the starting point for any model railway that is to be successful in giving you hours of pleasure and usage, not to say pleasure in construction, must be a clear concept of what you require from your model, and what it is to represent. Otherwise model railways can just grow in an ad hoc fashion which may cause problems and ultimately lead to loss of enjoyment and perhaps dissatisfaction with the hobby. After all, isn't the reason we take up railway modelling for enjoyment and relaxation? Nothing is more guaranteed to put *homo sapiens* off railway modelling than the frustration of a layout that cannot be operated as was hoped, whatever the reason, be it inoperable layout plans, warped baseboards or a general feeling that it is just not right!

I won't waste time and space on the choices of scale and gauge, type of layout or the respective merits of various railway companies and systems. Much as been written on these aspects already. However, I will start by offering some typical examples of styles and plans that exemplify different practices and reflect

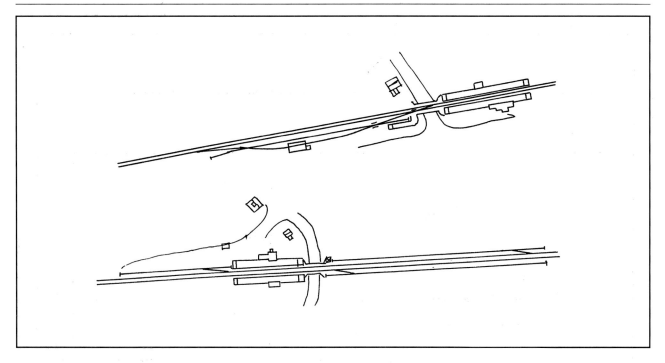

These two drawings show two station layouts from the famous Settle and Carlisle line, and are included to illustrate the style of a particular company on a particular route.

The Settle and Carlisle was unusual in Britain, being conceived and built as one entity rather than a hotchpotch of amalgamations, take-overs and extensions. Accordingly, it displays over all its length a consistent style both in the architecture and track layouts.

Obviously, all stations are designed to deal with the traffic they are expected to handle; however, on the Settle and Carlisle there was a further requirement, the provision of laybys. The smaller stations had minimum siding accommodation

and laybys as shown in the lower illustration, which is a sketch of the layout at Ormside. Those stations that were provided with covered goods facilities had larger goods yard layouts. Where a goods shed was provided, it was invariably located on a double-ended loop off a siding. Loading banks and cattle docks were normally provided as near as possible to the station building, thus necessitating only one access. The top sketch is of Long Marton and shows the minimal arrangement when a goods shed was provided.

Both sketches show the avoidance of facing pointwork on the main line - a feature only changed when modifications were made by British Railways on the ex-Midland main lines.

very clearly the origins of the prototypes, together with a few fundamental comments on basic considerations in planning a layout.

What type of layout?

One of the questions I have been most often asked when showing model railway layouts over the years at various exhibitions is words to the effect of 'Where did you get that plan or idea for the layout?', usually followed by questions on the overall size of the layout and its construction.

Conversation and even sometimes correspondence ensues on the lines of 'I have a spare wall, room, shed or garage and I have been considering building a layout. I want it to be as realistic and interesting as possible. How do you get the realistic look. . .?'

The short answer is to look at the real thing, contemporary or historic, identify and sift out what gives that railway or location its characteristics and re-create the impression of what you see and discover. I hope

what follows will be a practical and helpful response to that area of questioning.

The first and most fundamental decision to be made is what type of layout is to be built. The would-be builder must decide what he wants - loco shed, goods depot, main-line terminus or bucolic branch line, or even perhaps a preserved railway. Without that first fundamental decision, all subsequent organised action and development will be negated. Progress, if any, will be piecemeal, with the ad hoc growth and all the pitfalls referred to earlier.

Before building any layout I agonise on this decision, working out mentally themes and ideas, perhaps picking up inspiration from pictures or from sites that may have been passed while travelling, from articles or books or a particular railway line, route or station.

The next stage is to try to create a mental picture of how the finished model could look. If the family see me with a blank look on my face, inattentive, in a daze, then suddenly attacking scraps of paper with great vigour, pen in hand, they know I'm in the stages of early layout idea gestation!

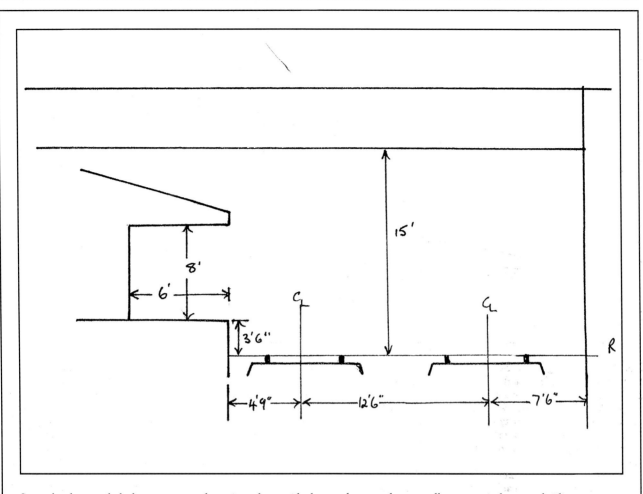

Some fundamentals before you start dreaming of your ideal layout! This drawing sets out minimum distances for clearances, assuming straight track; they should be judiciously widened for curved platforms, etc. All dimensions are in feet and inches.

To give an idea of some key measurements to be borne in mind when layout planning, here are some rough guides for O gauge:

Peco Streamline point	length 16³⁄₈ in
Peco Y point	length 15 in
A6 crossing angle point	length 17¹⁄₂ in
Model BR Mk 1 64-foot coach	length over buffers 18 in
Model of average six-wheel coach	length over buffers 10 in
Model of average four-wheel wagon	length over buffers 5 in
Model of small 0-6-0 tank, eg SR 'P' Class	length 7 in
Model of 0-6-0 tender loco, eg LNER 'J6' Class	length 14¹⁄₂ in

Remember that the length of platforms or sidings should be in relation to the length of the fiddle yard - it is no use having platforms and run-round loop that will take six coaches if your fiddle yard will only take two!

If your layout has a head shunt, don't forget to make it long enough to take the largest loco you intend to use. Also make sure that you allow space in loops and sidings to cover for the dead area at the end where vehicles cannot be left without fouling adjacent roads.

Do not expect large six-coupled locos and long-wheelbase vehicles to negotiate tight curves on a dockside layout. Remember that the 6-foot (O gauge) and 3-foot (4 mm) ideal minimum radii are much tighter than the norm on the real thing. The minimum radius allowed by British Railways for freight traffic is equivalent to 6 feet in O, while for passenger trains it would be 15 feet without a check rail.

Space can be saved in track laying using alternative formations. One three-way point will take up a lot less space than doing the same job with two standard points, and a double slip less than the two points it replaces.

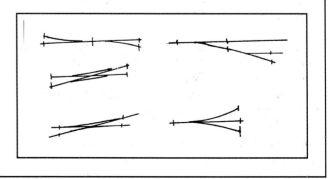

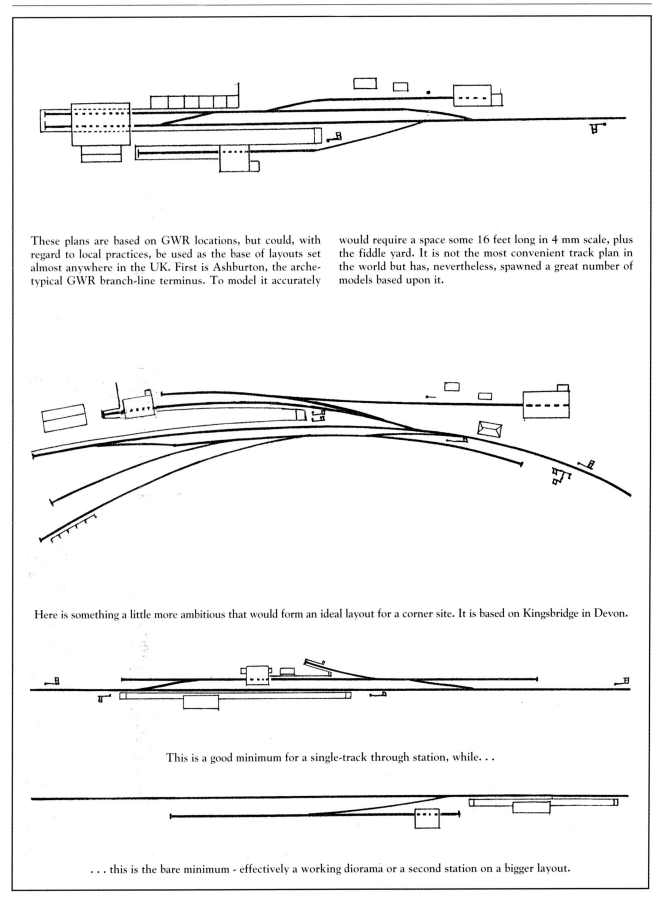

These plans are based on GWR locations, but could, with regard to local practices, be used as the base of layouts set almost anywhere in the UK. First is Ashburton, the archetypical GWR branch-line terminus. To model it accurately would require a space some 16 feet long in 4 mm scale, plus the fiddle yard. It is not the most convenient track plan in the world but has, nevertheless, spawned a great number of models based upon it.

Here is something a little more ambitious that would form an ideal layout for a corner site. It is based on Kingsbridge in Devon.

This is a good minimum for a single-track through station, while. . .

. . . this is the bare minimum - effectively a working diorama or a second station on a bigger layout.

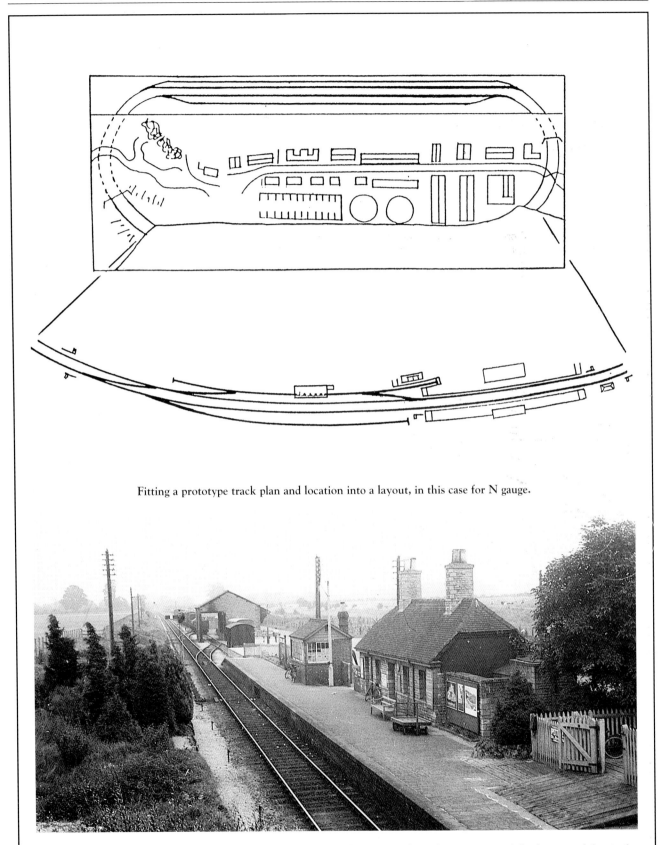

Fitting a prototype track plan and location into a layout, in this case for N gauge.

Lechlade station on the ex-GWR Fairford branch in Oxfordshire/Gloucestershire, showing many of the features of the single-track through station sketched opposite. *Lens of Sutton*

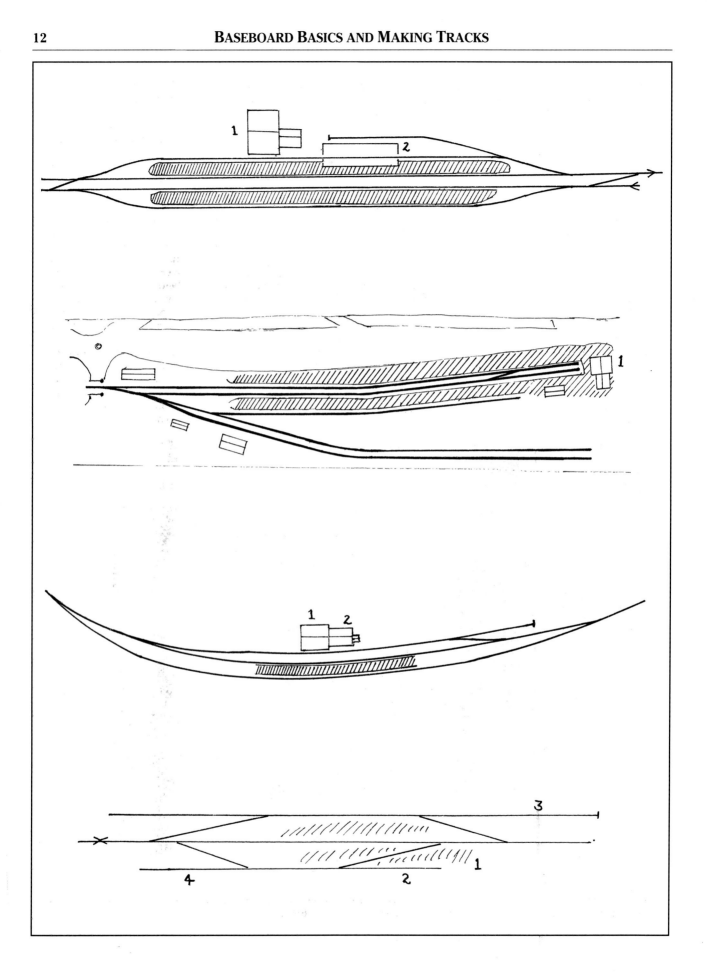

Left If you are looking for something a bit different, or even a side line from your main layout project, don't ignore the French or German scene. Treated with the same care in HO or N, and with the same philosophy offered in this volume, an excellent and convincing model can result. An advantage of choosing a French or German prototype is the very high-quality, reasonably priced, ready-to-run models that are readily available in the UK, but perhaps the biggest advantage of all is the quality of building kits and other structures available, far better than anything we have currently for the British scene!

The top diagram here is typical of a secondary station on a French double-track main line. It can represent any period from the steam era to modern-day electric and, coupled to a fiddle yard as part of a continuous run, would make an ideal layout in HO or N for some of the exquisite French models available ready-to-run in those scales. There are also excellent models of typical French buildings from Jouef in HO and MKD in HO and N, but be careful because the architecture has to suit the bit of France you are modelling just as much as the trains - what applies to UK prototypes also applies to those whose inspiration comes from further afield! This plan is based on Margival on the Paris-Laon route on the former CFN.

The second plan sketched is for a modern-image, contemporary suburban terminal and is based on St Gilles-Croix-de-Vie. It is again an ideal vehicle for excellent models available ready-to-run and could be accommodated in a space of 18 ft x 6 ft plus fiddle yard in N gauge. The station is simplicity itself and contrasts with our project layout. The area immediately behind the platform is a street scene.

The third drawing shows a typical single-track through station, based on St Victurmiem on the line between Angouleme and Limoges.

The platforms are low, barely above the height of the rails, and passengers walk across from the station building.

The final sketch is an example of a German track plan. Again this is for a small layouts, typical of German practice and designed to take advantage of the many excellent readily available models.

A final comment - given the excellence of many of the continental models, they almost demand proper use on a layout in the manner of the approach suggested with our British project layout. You don't need a field trip to France or Germany, however - although it is a good excuse! - since magazines such as *Loco-Revue* are easily obtained here (via Peco Publications, publisher of *Railway Modeller*) and you don't need to be fluent in French or German - a bit of patience and a dictionary will fill in the gaps that aren't obvious! Similarly there have been good English language books such as *French Steam* published by Ian Allan and *Steam in West Germany* (OPC), not to mention a host of pamphlets and small books and articles in the late *Model Railway Constructor*. There are also active societies in the UK for modellers of French, German and Swiss railways who are helpful, have knowledgeable members and often have access to libraries of books and magazines.

1 Station building
2 Goods shed
3 Loco shed
4 Loading dock

A bucolic French scene in HO using standard readily available commercial equipment, a bit of imagination and, of course, reference to the real thing - a theme which will be repeated in these volumes.

The idea for the project layout, for example, came about by trying to find ways of meeting the rather unusual criteria set for it. It required to be

- portable, ie capable of exhibition as well as domestic use

- capable of being housed in a normal domestic situation

- capable of fairly intense operation

- able to cope with trains of a reasonable length

- able to accommodate medium-sized locomotives - 2-6-2/2-6-4 tanks and 2-6-0 and small 4-6-0 tender locos - with believable trains

- able to provide for extensive and varied scenic treatment

- easily adaptable to suit any period from the 1880s to the present day

- and a design that would be easily adaptable to any part of the UK or even further afield.

That I decided to build the layout in O gauge placed on the one hand a further restriction on design, that of space requirement. But at the same time it gave me the opportunity to try to maximise the greater 'presence' that 7 mm scale models have by comparison with their smaller counterparts. It also gave me the opportunity to incorporate a reasonable level of

Right Large loco and short train. Austerity 2-8-0 No 90408 with three four-wheelers and a brake at Stubbins on 4 July 1963. *D. Hampson*

Below Large locos and short trains were not only the preserve of freight operations towards the end of steam. Here Stanier Class '5' 4-6-0 No 44686 with a three-coach set makes up the 13.28 Preston-Southport-Much Hoole service on 23 August 1964.

The close study of photographs is often quite revealing. The use of standard vehicles in sets was, despite plans and intentions, a rare occurrence until well into the 1960s. It was quite common to find local trains in the North West from the late 1930s to the late 1950s made up of motley col-

lections of coaches from various pre-Grouping LMS constituent companies. A formation would commonly be comprised of L&Y, LNWR and LMS-built vehicles; few trains were made up of the standard designs and sets beloved of modellers. *D. Hampson*

detail without being over fussy. Furthermore, it provided a challenge in developing something that hadn't been done before.

Looking at the criteria, I was taken first to two possible choices. The first was a pre-Grouping era setting using four-wheel and six-wheel coaches and short bogie coaches of the period, which would give a greater impression of space for a given platform length. (In the same way, a more modern era could be represented by push-pull sets or diesel multiple units.) The pre-First World War years of the pre-Grouping era combined, in what may be called the golden age of our railways, the advantages of short coaching stock, comparatively small locomotives - even the 4-6-0s of the time - and the elegance and grace of the Edwardian period.

The second choice was a period I remember, the 1960s, the end of steam, when all manner of strange operations could be seen; former express locomotives on trip work with short trains, and an air of decrepitude to contrast with the shiny elegance of the pre-First World War period. Decisions, decisions!

I looked further at the latter option, largely I think because it was what I remembered from my childhood and it gave more variety in terms of the type of stock and locomotives that could be run. There was something sad but fascinating about a decrepit 'Black 5' on two coaches, a 'Jubilee' on a short parcels, let alone a 'Britannia' on a ballast train! Similarly, the DMUs were becoming well established and the odd diesel loco could be interspersed with the steam. It was a picture that could to a large extent, by changing the locos and location, be representative of any of the regions of British Railways as it then was.

My thoughts next turned to the type of facilities that could be represented on the model, and in particular the creation of a small shed scheme or some kind of freight facility. This latter idea was given credence by my memories of Stanier '8F' 2-8-0s and 'WD' 2-8-0s shunting goods facilities with a handful of wagons. I also recall a 'Jubilee' with a 50 ft Full Brake and two fitted box vans shunting sidings used by a mail order company!

The idea I had for a shed layout was to produce a model of a small depot with turntable, coaling facilities and sidings to store odd coaches and wagons. There are plenty of books on engine sheds describing in detail buildings, fittings, operations and layouts. While recognising that sheds take up considerable space with these facilities and there are therefore difficulties in modelling a particular location, there is enough written about the sheds of the various companies and their operation to enable the modeller to develop his own along the lines of established

company practice. After all, having regard to the needs that the shed must serve, the railway companies themselves, unless having good reason to do otherwise, used standard layouts and designs for structures and fitted them in the space they had available, ie the land owned or suitable for acquisition.

By way of example, I have included overleaf some notes on GWR engine sheds and possible layouts, and on the modernised facilities and concerns the LMS put forward to rationalise and modernise its facilities to improve efficiency.

Thoughts on goods facilities were initially provoked by the site of a yard just outside Lyon in France, which could almost have been a model railway design, and a very similar one, but slightly less convenient for operating, seen in England, alas I

This scene of a small Midland Railway shed modelled in EM by George Martin shows what can be achieved in a small space. The layout allows locomotives to arrive on shed from 'off-stage' sidings (accessed via the bridge centre left) to be serviced and either stored on shed or sent out for their next turn of duty. This layout was very popular on the exhibition circuit and won the builder numerous awards. The photo, alas, does not do justice to the excellent fine detail work it contained, but does give a flavour. The arrangement lends itself to the 'kick back' style of fiddle yard described elsewhere.

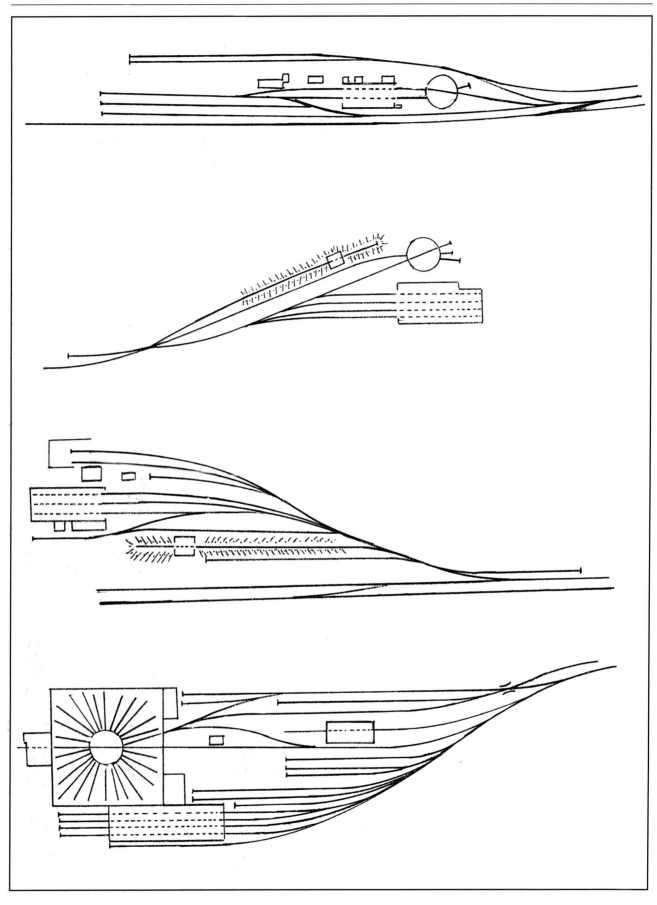

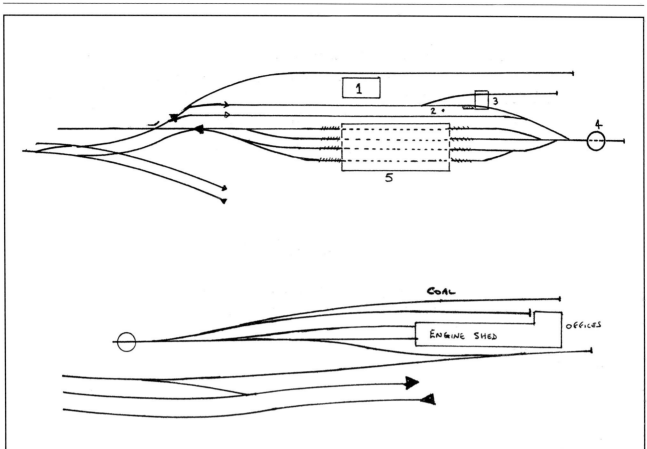

Left Some shed layout ideas based on a GWR practice, representing the different eras in which they were built.

The first is an early layout with the turntable immediately outside the shed; the second is a straight road shed with turntable and coaling stage; the third is a straight road shed without turntable; and finally a large facility with a 'turntable shed' and ancillary repair facilities.

Above The 'ideal' layout for motive power depots developed by the LMS in the mid-1930s to reduce the disposal time for locomotives. The numbers relate to the sequence of operations and facilities which gave minimum time for the necessary duties. The sequence was as follows: 1 Coal, 2 Water (the preferred option was to carry out these two operations simultaneously), 3 Ash-pit, 4 Turntable, 5 Shed, where the loco could be stabled or return to the pit for preparation for the next duty.

This 'ideal' layout was only one element in a number of factors that improved performance on the LMS. Those wishing to model an imaginary LMS shed or British Railways London Midland facility could do worse than use this as a basis for their layout plan, providing as it does all the likely facilities required by a modeller.

Alternatively, there is quite a large amount of published information covering locomotive sheds, many of which give detailed drawings of structures and equipment such as water columns, as well as track plans. Whether you want a 'stand alone' shed model or a smaller facility, you will find these publications an invaluable source.

By contrast, the second plan is of the ex-Midland and North Eastern shed at Ilkley, typical of a small shed. It was at one time home to eight or nine small tank and 2-4-0 tender locos, and provides plenty of scope in a small space for those who like building or collecting locomotives.

Sheds often provide the basis for preservation centres and, if you just enjoy collecting or building models of all kinds of rolling-stock, this could form the basis of a satisfying layout for you on which you could show your growing collection.

The small ex-GWR shed at Brecon in the late 1950s showing a good array of small locos on a site that could easily be reproduced in model form. Note also the assortment of small outbuildings, and the water tower on the extreme right. *Author's collection*

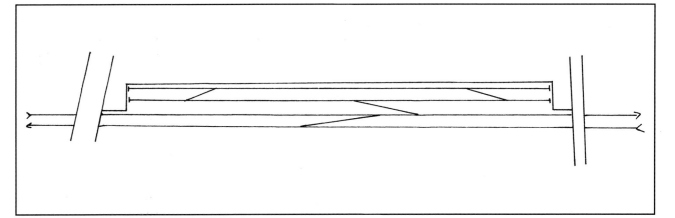

The sketch represents the warehouse sidings adjacent to a main line seen in similar configuration both in the UK and in France. The length is adaptable within reason to the space available and could provide an interesting shunting layout. Coupled to an adjacent fiddle yard, its scope is considerably enhanced. The two road overbridges - one could even be a rail overbridge or even a canal - provide a natural break and, of course, the platform area would be bounded by a warehouse or series of warehouses giving an ideal backscene or frame to the layout.

This basic idea, omitting the main line and replacing it with sidings and finishing the frontage as a dockside, would provide a basic dockside layout. There are many excellent ship model kits of vessels from sail to modern times, some of which are at, or very close to, the modelling scales. For example, Caldercraft have kits of a 1914 vintage coastal tramp in 1/48 scale, ideal for O gauge. However, these kits are a hobby in themselves, not just as quickly made scenic adjuncts.

cannot remember where. Both these facilities served warehouses, possibly postal distribution points, and comprised parallel sidings adjacent to main running lines. In both cases, but particularly the French one, the sidings ended conveniently at bridge abutments carrying roadways over the railway, thus providing the sort of ideal scenic break dreamed of by modellers of prototype locations. Again, photographs of many locations, particularly in the north of England, show large locomotives shunting a few wagons around old goods warehouses amidst weed-strewn tracks.

Researching the prototype

It is amazing how we forget what is on our own doorstep! A look through the local model railway club photo archives revealed a facility that, if it were drawn as a plan or shown as a model, may well have engendered disbelief. However, I became rather taken with it and still believe it could form the basis of an excellent working model.

I've included some pictures of this facility and the line that served it. It was a coal yard, effectively a fan of three sidings off a single-track goods line that once served collieries and local industry away from a town centre and its major rail facilities, but still in an urban setting. The thought of this little facility, its enclosed surroundings of terraced houses and the '8F' shunting its train of three 16-ton mineral wagons was almost enough to send me running for timber,

screws and the beginnings of a new layout!

A similar possible prototype for adapting to model form was also uncovered, again locally. Astley Bridge was almost the archetypal model railway branch-line terminus, neat and compact, the station area entered on a curve, the whole lot set in a shallow cutting bounded by stone retaining walls. The line had in truth been built as such, but became freight only after a few years' existence, passengers preferring the more frequent and cheaper tram! It survived as a goods yard just until the end of steam, I think, although I've never seen evidence that it was very busy. I recall it being one of the last haunts of the Lancashire & Yorkshire Railway Aspinall saddle tanks in the early 1960s.

These are just a few examples of sources of inspiration for a model railway that I found locally, and clearly there are many, many more. In fact, you would be unlucky if you couldn't find something near you or in the area you are interested in modelling.

There must have been several small yards, for example, in the London area, operated by the major railway companies, many of whom worked directly or indirectly into the capital. What about mineral railways in Cornwall, or the small quayside railways that would have been seen around our coastline? The rural sidings of East Anglia, the railways of the coal mines? The possibilities are there all over the country if you take the trouble to seek them out.

Don't forget the many wonderful photo albums, line histories and general railway books that can often throw up photographs showing interesting possibilities for modelling. The later books which get away from the standard three-quarter front view of express trains are often a good source, as are the historic ones. They can also provide cameos and details that cry out to be modelled - maybe not on this layout, but the next. . .? You'll quickly find that you develop an eye for these types of pictures and lay them mentally or literally aside as potential sources of information or modelling inspiration.

Modelling an actual prototype to scale can be a very rewarding process and requires patience and great care in seeking out factual information on your chosen location. However, a short cut would be to pick a prototype that has been covered by one of the many high-quality line histories that are available, thus cutting out most of the leg work in researching the details needed to build an accurate representation.

The research in itself can be both time-consuming and rewarding, and may lead to the provision of much useful information. BR, the National Railway Museum, specialist societies, local libraries and record offices, local newspapers and advertising both locally and in the national railway and model press, are essential routes to the acquisition of the necessary information.

Another alternative is to take your favourite prototype and, if you cannot find the space to model a specific location or feel that this would be too restrictive, look for plausible extensions, branches or additions that you could develop to fit the space you have and to suit your own requirements, 'in the style of', as they say in the antiques trade.

One local source of inspiration was the Astley Bridge branch. The layout plan is based on the station as it opened when it had passenger facilities; however, it had a short life as a passenger facility, being opened by the Lancashire & Yorkshire Railway in 1877 but losing its passenger facilities in 1879! It remained in use as a freight-only facility until 1961 and until 1964 as storage for condemned vehicles. It is now the site of a drive-in burger joint - aargh!

Again it suggests a nice simple self-contained model set in a cutting with substantial stone retaining walls all round. Behind the station entrance is terraced property, while there is a separate entrance to the goods yard proper from the main road at the bottom of the plan. This access in stone sets dropped in two stages to ease the gradient and make haulage up to the main road level easier. All in all, a nice compact little model could result from this - a treat for those who like shunting and goods vehicles.

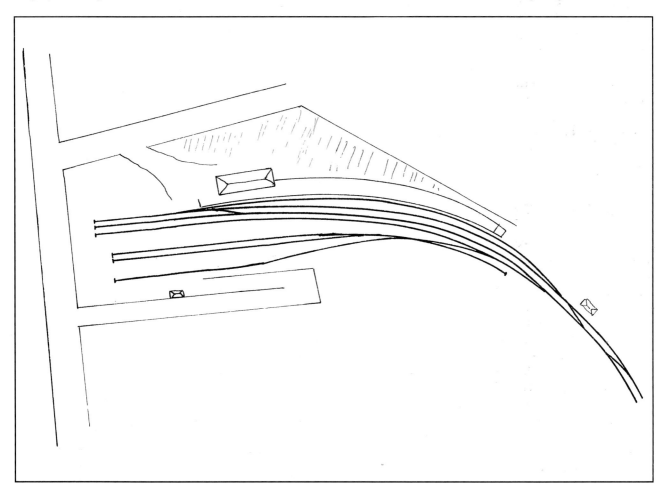

A plan of the coal depot in Bolton mentioned in the text, known officially as Hulton Sidings, showing the layout of the track. In fact, the site was only built on in 1992, becoming a supermarket - yes, you've guessed, just as I discovered it!

Situated on the original Bolton & Leigh Railway of 1828, the yard is surrounded by terraced property and brick buildings such as stores and offices, and the railway is in a cutting just past Daubhill Junction. Having made a comment elsewhere in the text about the incongruity of a half-timbered pub at the end of a Victorian terrace, here is almost an exception, a pub, the Stag's Head, which is in a sort of Edwardian Tudor, unusual but not unique in this location, and with its own cobbled forecourt setting.

I would envisage the yard being a basis for a working diorama rather than a fully fledged model railway, but one with which you could have a lot of fun and, if done to reflect its surroundings, would take quite a time. I don't know when the coal conveyor was installed but it doesn't show up on plans of the yard I've seen dated 1953, which still show the sidings where they are located in the diagram.

It was not uncommon for traffic across one of the main A road approaches to be delayed as the yard was shunted. The late '50s and early '60s often saw an '8F', a couple of 16-ton mineral wagons and a brake van! This could prove a rather attractive proposition in any scale using either the Graham Farish, Hornby or CCW '8F' kit in N to O gauge respectively. One day. . .?

To make the model more easily operable I would couple it to a small fiddle yard and, with modellers' licence, add a loop to the 'main line' to make shunting easier! Fitting this layout into a small space whatever gauge you choose can be helped by using the fiddle yard as the closing part of the loop. I think my preference for the location of the fiddle yard would be beyond the road overbridge; I would straighten out the 'main' line and add the loop on the opposite side from the yard itself, ie the cutting side. Were there space available, I would also model the level crossing. Space could be saved by making the fiddle yard a traverser or sector plate type with the dual purpose of closing the loop. This arrangement is sketched elsewhere.

Above right A view of Hulton Sidings in around 1960. This little scene provides a cameo not only of a railway age long gone but also a social age. This site is now a supermarket. What inspiration for a model!
D. Hampson

Right A typical late era coal train holding up traffic on the ungated crossing outside Hulton Sidings, which are just to the right behind the tender of the '8F'. *D. Hampson*

Right
Another freight backwater - an '8F' and a few goods wagons in an urban coal yard. Rudimentary track, cobbled roadway, piles of coal and general industrial paraphernalia, all contained in a long narrow site, ideal for a small-space model.
D. Hampson

Left Photographs are an invaluable source of modelling inspiration. This is Blackrod Station in 1947 with an ex-Lancashire & Yorkshire railmotor in the platform. Note the absence of clutter on the platform and the LMS tubular steel signal post. A sharp contrast to the L&Y notice behind the wooden fence. I wonder if the bag on the platform belonged to the photographer - today the sight of such an object would no doubt be the cause of considerable alarm.

Below and opposite Bacup station, East Lancashire. The first view (*below*), from July 1963, shows similarities with Bolton's Great Moor Street, inspiration for Platt Lane. With the closely abutting buildings, it is a further useful source of inspiration for a compact model. *Above right* The platform end - note the clutter of lamps, signals and water crane, together with evidence of recent ballasting, and mixed flat bottom and bullhead rail. The third view (*below right*) shows the station in its final days, with diesel units and weeds the order of the day. Again it provokes thoughts of how this could be adopted to model form. *D. Hampson*

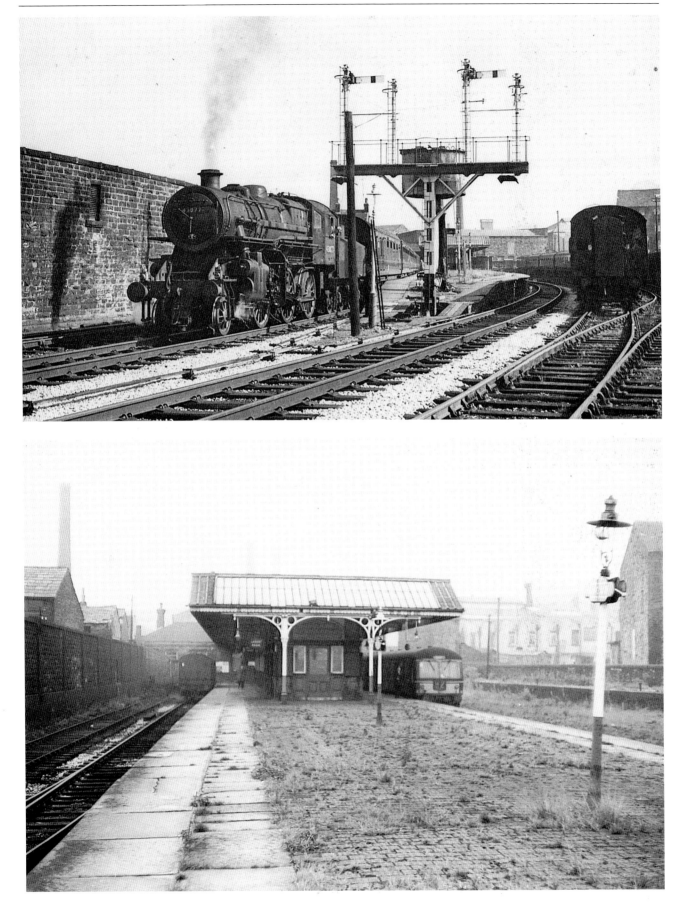

Opposite above Station buildings come in a bewildering variety of shapes, sizes and architectural styles. Only a photograph proves that Lostock Junction Station booking hall carried a water tank above it! This booking hall was at road level with steps leading down to an island platform. BR corporate ownership is well established by 1959 when this picture was taken. *D. Hampson*

Opposite below There's a wealth of detail here, with LMS-origin hut, colour light signals and details of track maintenance paraphernalia stored between tracks. In addition, the skyline shows an interesting arrangement of buildings. The year is 1963. *D. Hampson*

At the other end of the country, and quite a different setting, here's the terminus of the Faringdon branch in the 1950s, seen from the road approach side - a scene beloved of railway modellers!

Usually modelled in the kind of space allowed for Platt Lane, such stations actually occupied large sites, since rural land was relatively cheap, hence the 'openness' of the scene apparent in this view.

Another ex-GWR terminus, this time Kingsbridge, in 1962. Again, a great deal of detail is available to the small-space modeller, including the compact diesel unit. *Both author's collection*

This adoption of a 'might-have-been' line extension is something I have done and found very rewarding. The first O gauge layout I built was intended as a minimum space trial - a trial really of scenic techniques as much as anything. The whole thing eventually measured only 8 ft x 2 ft 6 in, and was a classic example of what not to do in the sense that the scenic test track became a very popular and successful exhibition layout in the two years that I owned it. In that process it became adapted to better suit its unintentional fame and was a classic example of the piecemeal development I warned you about earlier. Nonetheless, despite all its failings from lack of real design and its small size, I think it still gave me far more pleasure than any other layout I've owned. I often wonder what became of it after it was sold on after my short ownership.

I had wanted to build something to represent the Kent & East Sussex Railway under British Railways but alas, none of the actual stations proved suitable. I therefore came up with a plausi-ble extension to Dallington, or possibly a Colonel Stephens-engineered branch off the Hastings line. Both were highly improbable but, in that crazy age, just possible. The station building - the only building on the layout - was a model of Northiam on the K&ESR and was a typical Colonel Stephens corrugated iron affair. The width of the model was actually reduced 25 mm in width as the cornflake-packet mock-up looked much too big when placed on the layout. I even had a plausible excuse for a modernish bridge carrying the road over the railway conveniently before (and hiding) the entrance to the fiddle yard - the build up to D Day required many developments to the infrastructure in that part of England, including the replacement of the more normal level crossing at Dallington Road! The war also gave me the excuse to model a smaller WD store of the type located at

Above right Another view of Dallington Road, from the buffer stops.

Below right Ex-LB&SCR 'Terrier' at Dallington Road. This locomotive was built from a high-quality white-metal kit and provided sterling service on the layout - it is now resident in Belgium. Careful inspection of the track will reveal how convincing the ready-to-lay O gauge track can be, carefully laid and painted.

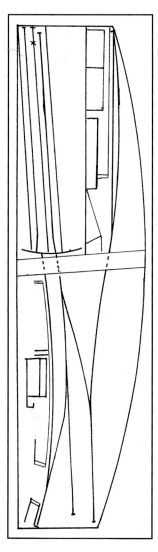

Left Dallington Road, the minimum space O gauge layout referred to in the text, occupied a space just 8 ft x 2 ft 6 in. The bow front was a late addition sloping down from rail level and helped create the illusion of a little more space.

The sector plate not only provided the fiddle yard in a minimum space but also provided the closing of the loop and was thus an essential element to shunting, necessitating one fiddle yard road, or at least a goodly proportion of it, always being kept empty to allow locomotives to run round, etc. This same type of convenience is used in the Platt Lane project to save space.

Right Dallington Road. Even on such a minimum space layout it is possible to create an authentic atmosphere. It employed ready-to-use Peco Trackwork, as used in the project layout Platt Lane, which can be made to look very effective. Despite being an O gauge layout, Dallington Road used a lot of bits and pieces intended for 4 mm scale!

Tenterden on the K&ESR, but by now in peaceful use by a timber merchant! Modeller's licence! Judging by the reception the layout got, it must have succeeded in representing the Kent & East Sussex.

Returning to the project layout, some of the photographs I came across at the same time as those of the coal yard were views of a now long-gone station known as Great Moor Street in Bolton, gave me the idea. Despite this being very much the second station in the town, an LNWR incursion into the Lancashire & Yorkshire heartland, it would still have required an awful lot of space to model it, even if I had wanted to. I didn't, but the photographs did reveal a number of interesting features that gave me the inspiration to think of further possibilities for a project layout to meet the criteria I had set.

By now I was into the possibility of a town terminus

with good facilities and able to accommodate three-coach trains, with some ideas drawn from the photographs for the scenic side and how the layout should look. These included a bridge at the platform ends carrying the railway over a roadway, a platform canopy, the railway above street level, a station building across the platform ends, and a setting of terraced houses, shops, factories, etc, to represent an industrial town.

A few sketches gave me an idea of what I wanted visually, and sketches of possible track plans followed. These working sketches, crude as they are, are included as they show that not only am I unable to draw, but also that the initial wave of enthusiasm has to be tempered by what is actually possible. The sketches eventually came down to a layout that met the criteria and could be fitted into a garage (ie 16 ft x 3 ft) if nowhere else, and Platt Lane was under way!

Opposite The inspiration for the project layout, a sad and decrepit Great Moor Street station, Bolton, awaiting demolition. Great Moor Street was very much the LNWR's second fiddle to the Lancashire & Yorkshire Railway's Trinity Street station in the town. Apart from summer trains to Wales, when a Patricroft 'Precursor' 4-4-0 might provide welcome relief from the local trains operated by Plodder Lane shed men, nothing very exciting seemed to occur. *D. Hampson*

Below The exterior of the main station building awaiting its fate in 1962; the train shed can be seen behind the main building. The building in its earlier days sported a canopy, the shape of which can be determined from the brickwork on the front. The door and window surrounds were in red, white and blue brick. The station's only use by 1962 seems as an advertising space and parking lot for vehicles under repair. *D. Hampson*

A look up at the side of the train shed and the road access to platform level. *D. Hampson*

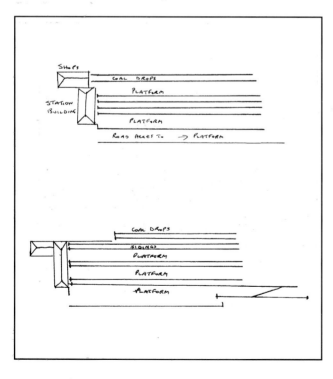

Two sketches showing the basic arrangement for platform and coal roads at Great Moor Street. The top one shows the arrangements in drawings dated 1869 for the 'proposed new passenger station at Bolton'. There were two platform roads between which were two sidings; whether or not loco release facilities were proposed is not clear. There were two parallel coal drops roads immediately behind the wall supporting the overall roof, shown as 'coal viaduct for Mr Hulton's coal'. The width of the coal facility from the edge of the platform wall to the outer edge was shown as measuring 24 feet and covering the two roads only. The roof was a single span measuring 82 ft 8 in across and standing in the centre 41 feet above track level. The platforms were 20 feet wide and 2 ft 9 in high from rail level.

The lower plan shows how things were actually built with two sidings between the train shed and the coal drops and an overall roof in two spans. An object exercise in the differences which even official drawings show!

Right Some doodles to show how I felt the prototype information could be translated to a model railway layout plan - part of the gestation of Platt Lane.

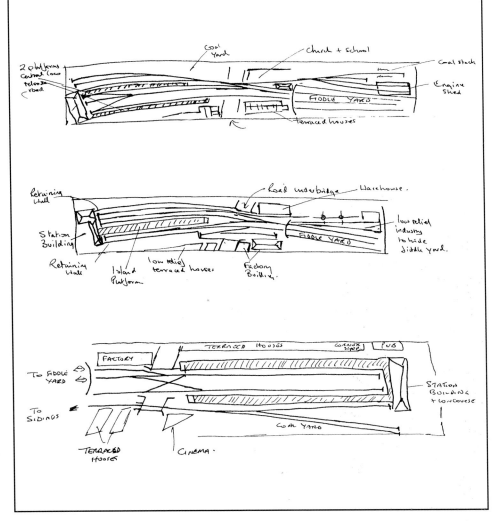

Below The final plan for Platt Lane.

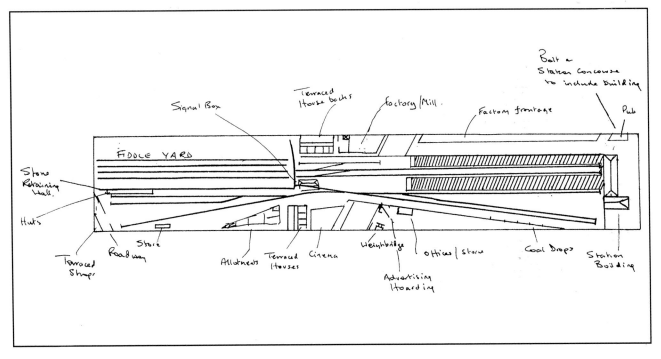

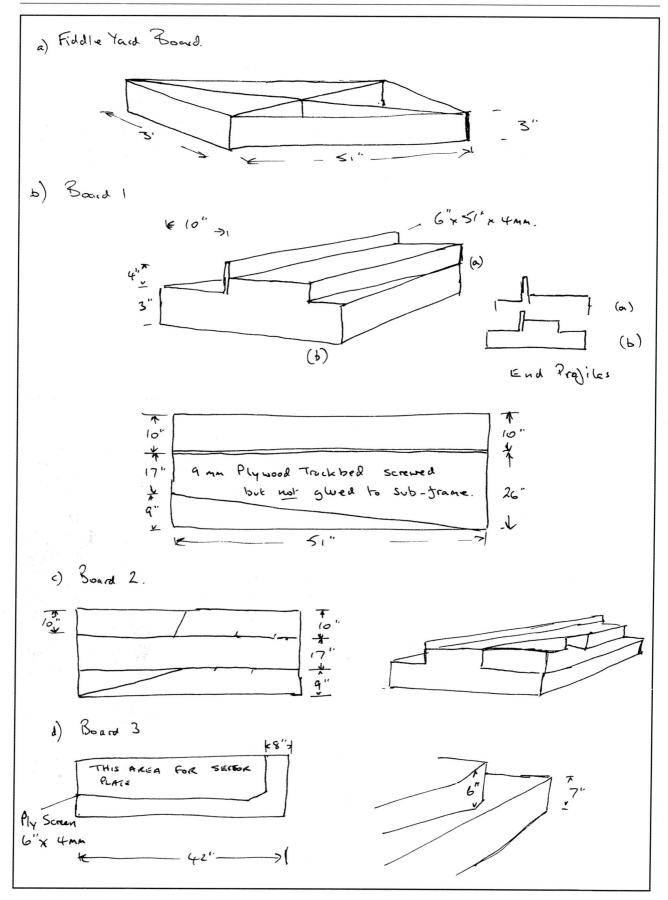

a) Fiddle Yard Board.

3'   3"   51"

b) Board 1

10"   6" x 51" x 4mm.

4"   (a)

3"   (b)

(b)   End Profiles

10"   10"

17"   9mm Plywood Trackbed screwed   26"
but not glued to sub-frame.

9"

51"

c) Board 2.

10"   10"

17"

9"

d) Board 3

8"

THIS AREA FOR SECTOR PLATE

Ply Screen   6"   7"
6" x 4mm

42"

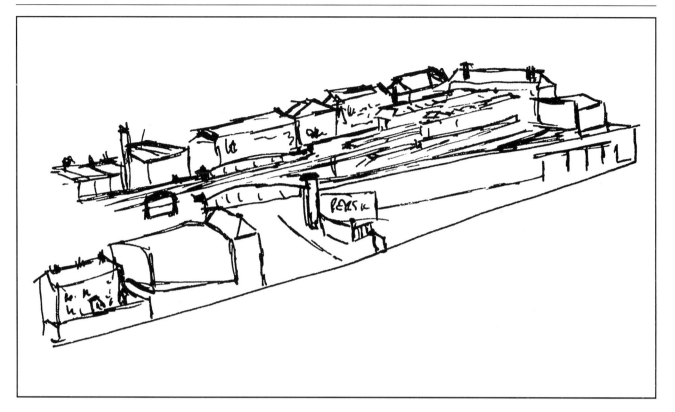

Left Preliminary sketches showing the evolution of the Platt Lane baseboard and its different levels.

Above A thumbnail sketch providing a three-dimensional 'fleshing out' of the plans to give some idea of what the finished layout might look like.

Below Mock-ups of the terraced house backs in place on the layout, together with loco and stock of the type we will be using, all helping to get a 'feel' for the model. The scenic development of Platt Lane is dealt with in detail in the next volume.

2.
PROTOTYPE TO MODEL

Marking out and mocking up

Having made a start on translating the prototype to model form, the design process now moves to the floor! However careful you are in drawing out your plan from the earlier sketches, you can't beat drawing it out full size. I use rolls of decorator's lining paper cut to just larger then the proposed layout, the width being made up by taping lengths parallel to each other. I then mark out the boundaries of the baseboard and divide the area into pencilled 1 foot squares. The proposed track plan can then be drawn out full size.

In the smaller scales, track templates are available for both straight track and curves to various radii. Most of the major scale track manufacturers will make available drawings, to full size, of the pointwork that they manufacture or that can be made from their components. These are invaluable and certainly worth the modest outlay to enable the position of pointwork to be assessed with some degree of accuracy.

Don't be surprised if at this stage you can't fit in the layout that you sketched earlier. Better to find out space problems now than at baseboard and track-laying stage!

This is also the time to work out the arrangements for platforms, structures and, very importantly, the location of the baseboard joints and arrangements, in the case of an end-to-end layout, for the fiddle yard.

Baseboard joints should be arranged to avoid pointwork and structures, or features across which any joint cannot easily be disguised, for example the middle of a bridge. Consideration also needs to be given to the size of the finished baseboard sections and how they will be transported. For example, 6 ft x 3 ft baseboards will fit in very few cars. The sketches in this chapter show where the baseboard joints were located for the project layout.

Using lining paper marked out with a 1 foot square grid as described in the text. Lengths of trackwork, odd buildings, boxes, locos and stock, whatever is available, can be used to given an impression of what the model will be like, and whether loops give sufficient clearance, etc, before the track plan is marked out on the paper. It is very important at this stage to mark baseboard joints, thus avoiding pointwork spanning them.

Baseboard design and construction is discussed later in detail, so suffice to say here that I believe the process needs to be considered as an integral part of the layout to gain maximum effect and optimum usage.

Buildings and structures form an important visual element of a model railway and their inter-relationship with the railway itself and the practicalities of their location in relation to their function also need to be considered - they must be located in relation to the function they perform.

Odd lengths of track, items of rolling-stock, boxes, books or, if you are really organised, cardboard mock-ups of the main buildings, can be moved around and adjusted at will until you get what you're looking for - or even something better that you hadn't expected!

The great advantage of the full-size lining paper work-out is that it gives you the opportunity to consider all these issues at your leisure before cutting the first timber or even committing expenditure to any extent on a layout.

One feature of Great Moor Street that I found rather unusual, and which became a feature I wanted to represent on the model, was the provision of coal drops. Not unusual, you may say - a common feature on North Eastern branch lines. Well, that's as may be, but in a large town centre, next to the passenger facilities and away from the goods areas?

This and the bridge carrying the railway over the

The development of a scheme for a model railway is dealt with in detail in the text. Indeed, this first volume is really about the design of a model railway and the construction of those essential but rather boring foundations on which ultimately the success of the layout depends. Having decided on the concept and worked out how the layout will look, I find, particularly with a layout that is multi-levelled, that it is a useful exercise to visualise the finished model in 3D. A simple scaled-down mock-up, crudely constructed from scrap card, will be sufficient. The scale chosen is usually one-twelfth, or 1 inch to 1 foot. It is surprising how useful these mock-ups are in giving a feel of the finished layout and providing a visualisation of your design. They also provide a testing ground for your baseboard construction ideas and can highlight problems and opportunities arising from even the most carefully drafted plan. Note, for example, how the angle of the inked-in road underbridge has been changed.

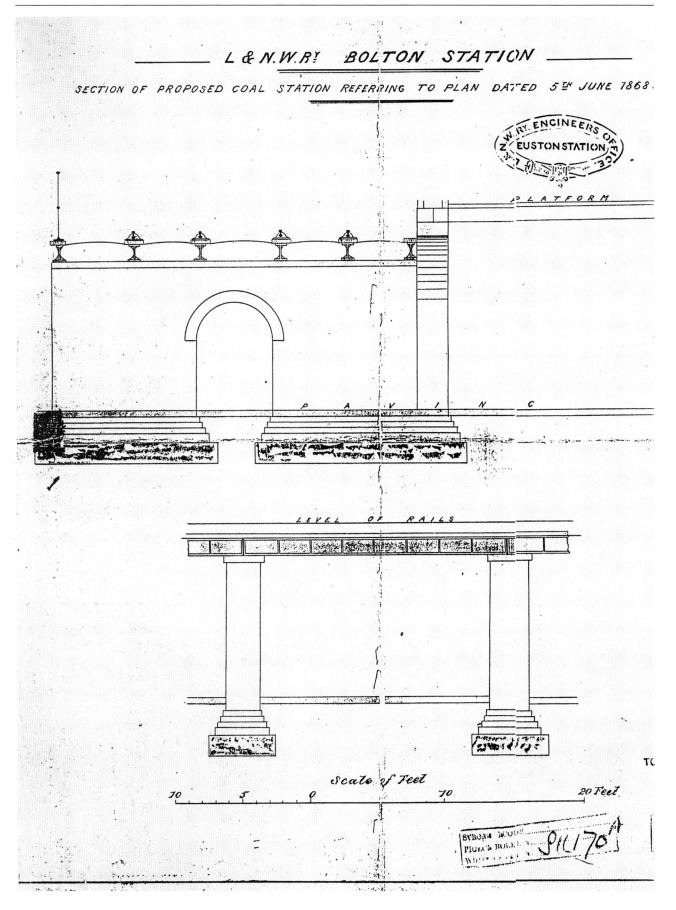

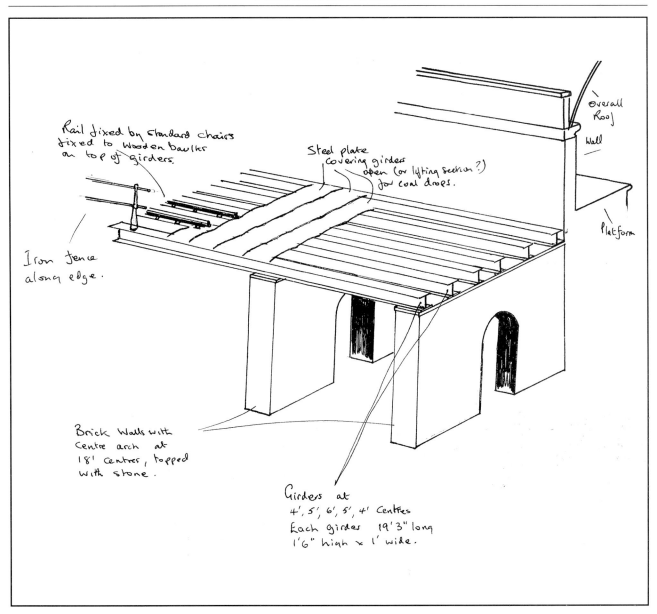

Rail fixed by standard chairs fixed to wooden baulks on top of girders.

Steel plate covering girders open (or lifting section?) for coal drops.

Overall Roof

Wall

Iron fence along edge.

Platform

Brick walls with centre arch at 18' centres, topped with stone.

Girders at 4', 5', 6', 5', 4' centres Each girder 19'3" long 1'6" high × 1' wide.

The original LNWR plans for the proposed coal drops at Great Moor Street, and a sketch from which the model will be prepared. A detailed description of their construction is given in a later volume.

roadway would necessitate a multi-level baseboard. Here again, time spent working out the design of the baseboards is time well spent.

The situation so far is that we have arrived at a track plan, and have decided largely what structures and scenic features are to be included, and thus the contours of the landscape. The baseboard sizes should also be clear by now.

I have found that at this stage, particularly as there will be a series of baseboards with varying levels which must all be right to create the effect sought, it is wise to construct a mock-up of the proposed model. Nothing elaborate is required - an old cardboard box cut up to represent the baseboard is all that is necessary. One-twelfth scale is big enough to find any pitfalls.

In the case of the project layout, the mock-up it highlighted two problems. First, I hadn't allowed enough space in the plan for the fiddle yard entrance, and second, I had more space than anticipated in the area between the road and the coal drops. The effort spent in producing the mock-up was already proving worth it.

The mock-up is illustrated on page 35 and is made in sections corresponding to the individual baseboards that make up the layout. The size of each section of the mock-up and the height of the coal drops, bridges and retaining walls need to be represented accurately to give the impression of the final layout. If it doesn't look right on the mock-up, change it. If a retaining wall is too high and looks out of proportion, try a smaller one. Don't forget to keep a note of the final sizes and dimensions with which you are satisfied, as these will be transferred to the actual layout. The trackwork can be marked in with felt-tip pen, and platforms indicated by lengths of cardboard laid flat.

I altered a couple of other things from the original plan; the angle of the road bridge was changed, and I straightened out the platforms, something I have since regretted from a visual point of view, but which at the height that the layout is normally viewed is not really noticeable.

I think most end-to-end layouts need to have provision to be operated from either side, the front at home, where invariably the layout is against a wall, and from the rear at exhibitions. This leads to a number of problems and, inevitably, compromise. It means that you have to consider what the layout will look like from both sides and finish it fully on each. The biggest problem, however, comes with disguising the fiddle yard.

Fiddle yards

The usual dodge of putting a scenic feature or railway extension in front of a fiddle yard, Bollings Yard Sidings in the case of Platt Lane, is fine, but it does mean that you have difficulty seeing what's going on over the fiddle yard screen if you are operating from the rear, and also in fundamental matters such as coupling and uncoupling; it is a very difficult act to reach over a screen to almost three feet away and assemble a train of vehicles fitted with three-link couplings! Similarly, if you are operating from the front, even with a conventional ladder fiddle yard with access via points, you will need to change ends with locomotives and brake-vans.

A simple answer would be not to disguise the fiddle yard but leave it open. This seems to be becoming an increasingly popular feature on exhibition layouts and gives the viewing public a chance to see the rolling-stock collection. However, in these times of security concerns, it also makes theft quite easy unless your fiddle yard end is manned at all times! I prefer to give exhibition viewers a surprise so that the casual viewer is, as on the real railway, not entirely sure what might be coming into view next. It is surprising, particularly with younger viewers, how much interest there is in guessing what the next loco will be!

With this layout it would be possible to leave the fiddle yard open; indeed, this would considerably ease construction. It would, however, detract from

Here are a few alternative arrangements for fiddle yards.

The first (a) represents the simplest - a fan of sidings from plain turnouts. Ideally, this should be repeated at the far end to enable locos to be detached from their trains and run round them without handling. Seldom, however, is there room for this luxury! This arrangement takes up a lot of space if you are going to get a decent length of siding.

In (b) substitution by three-way turnouts saves a bit of space, but these points are costly to buy, or take a great deal of time to build (at least for me).

Diagram (c) shows the arrangement used in the project layout and a favourite of mine. This is a simple sector plate swinging from a pivot, X. The ends of the fiddle yard roads need to be splayed out to ensure alignment with the entrance road when the sector plate is swung to use that road. Alignment is achieved by eye, and locking pins (see page 41) hold the road in place and provide electrical switching. This arrangement saves a great deal of space.

Diagram (d) is a variation, known as a traverser. Instead of a pivoted swinging fiddle yard, the whole lot slides back and forth. Alignment and electrical arrangements are made as in (c).

With both (c) and (d) the movement is aided by the liberal use of 'iron-on' melamine timber edging strips on the bearing surfaces - simple but very effective. This arrangement can also be used across curved track on a continuous-run layout (or indeed the straight section of a traditional oval). For use in a continuous-run circuit, simply repeat the same arrangements at both ends. The essential point is that the movement of the traverser must be straight; many ingenious devices using screw-threaded rods and handles have been devised to achieve this even movement and good alignment. Personally, two hands and a pair of eyes have always worked on my layouts.

Diagram (e) shows a 'kick back' arrangement where trains are reversed into and out of storage roads beneath or behind the main scenic portion of the layout via a single-track access road linking them to the main station or whatever. Even more space can be saved by the use of a sector plate. This type of arrangement is well suited to docks and other types of industrial layouts where the many buildings and structures required make hiding the fiddle yard easy.

Drawing (f) is a crude attempt to show how pinning and gluing a strip of ply, softwood or even hardboard to the edges of the fiddle yard deck helps prevent locos and stock descending to an undignified end on the floor - I know, I've done it! This arrangement should be repeated on all exposed edges, particularly the sides and ends. Simply fixing dowels in a parallel line - say, three or four pairs across the width of the deck - allows you to slot in a piece to protect the open end when trains are not coming in or out and the deck is being moved to align for the next train movement. For preference I reckon the height of these protective barriers should be not less than half the height of the locos and coaches to be used.

the operation and the effect of the layout. If more space were available, of course, the fiddle yard could be an 'add-on' and the baseboard developed scenically across its full width.

As I said earlier, compromise is always an integral

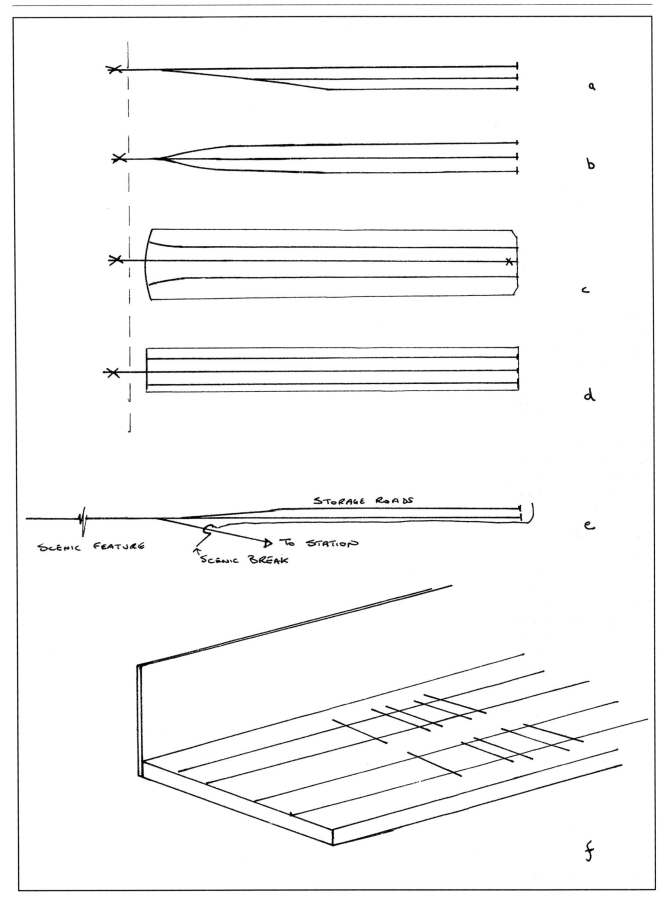

a

b

c

d

STORAGE ROADS

SCENIC FEATURE

SCENIC BREAK

TO STATION

e

f

part of railway modelling, and as the layout would see little use at home it was developed as planned, to be rear-operated. So far as the inconvenience of coupling and uncoupling the Bollings Yard Sidings is concerned, there are ways to overcome this. As designed, this section of the layout could be wired to be completely self-contained, and could be worked without interfering with access to the platforms. It could therefore be an ideal candidate for some form of automatic control and, with discreet automatic couplings on some goods stock, a simple but automatic shunting sequence could be operated, a useful feature perhaps for exhibition use or even to watch at home.

Before moving on, a few words on the fiddle yard itself might not go amiss. The fiddle yard, storage sidings or whatever you wish to call it, whether part of an end-to-end or a continuous-run layout, essentially represents the rest of the railway system; it is wherever trains depart to or arrive from, be they passenger, goods or locos for your shed yard. It provides the illusion of reality in that its sidings are generally either hidden or, by some means or other, clearly divorced from and not part of the main scenic part of the layout which is the focus of attention. To draw an analogy with the theatre, it is the off-stage part of a production.

There are many varieties of fiddle yard, the simplest of which is the ladder arrangement, where, on entering the 'off-stage' facility, the train runs into a series of consecutive points opening up to a number of dead-end sidings; two points open up into three sidings, three into four. In a continuous-run layout the sidings are closed at the other end by a similar arrangement of pointwork; this enables trains to leave and enter the sidings from either direction.

It is a simple and straightforward matter to arrange for remote control of pointwork. With the use of either an electrical dead end to sidings or a short section of loop switched off at one end, the fiddle yard can be controlled to a large extent from the main control area. Access is needed at the yard itself to break up and rearrange trains by hand.

The problems with these simple and reliable arrangements are basically twofold - they are wasteful of pointwork and the pointwork required takes up a lot of space. Some space can be saved if a three-way point is used, but there are other ways of building fiddle yards that are more economical in both regards!

A simple alternative to the use of points is the construction of a traverser. This uncomplicated device requires a bit more in the joinery and electrical departments but is a remarkably efficient means of dealing with the problem. It also has the advantage of being able to be used in a continuous-run situation and can bridge multi-tracked running lines. It can also be used in a curve.

In effect, a traverser simply enables you to slide a bank of running lines or sidings so that you can line up the one you wish to use with the appropriate running line(s) entering and leaving the fiddle yard. Traversers do away with the need for any pointwork and thus save a considerable length and make more space available for the actual trains themselves.

While many complex and ingenious mechanical arrangements have been developed over the years to ensure that they line up exactly, more swiftly and even automatically, unless you are into devising drive mechanisms the simplest arrangements, shown in the accompanying drawings, will be sufficient. In essence all that is required is to arrange for the baseboard carrying the sidings to slide, some means of preventing it being accidentally slid off the layout altogether, and a means of switching electrical contact from the running line to the siding in use.

I have found that the arrangement shown has proved more than adequate for my needs on various layouts: manual movement of the traverser, alignment by eye and the final alignment locked in place by a peg and socket of some type which also provides the electrical contact for the road to be used automatically.

One thought that occurs and deserves a mention while talking of fiddle yards, is the safety of the stock. I reckon that more damage occurs to rolling stock and locomotives in the fiddle yard than anywhere else. Not only does the stock suffer most handling there, but it is so easy to knock it over reaching across for something or other. Baggy jumpers are definitely out when operating a model railway!

Seriously, I know from bitter experience that you should always provide safety rails around the three edges of the fiddle yard. The one at the end provides a buffer and those along the side prevent the stock from accidentally falling off and on to the floor. Some builders line the inside of the safety rails with foam to protect their stock, but my experience has been that coach handles, butterflies and other protrusions get caught in the foam and are damaged. My preference is just to leave plain, clean wood.

A third option is the sector plate. This is similar to the traverser except that it pivots at one end, the other being swung round until the appropriate siding lines up with the road entering the fiddle yard; because it is pivoted at one end it is not suitable for use on a continuous-run circuit.

The sector plate arrangement is used on the project layout and its detailed construction is described subsequently in the section relating to baseboard

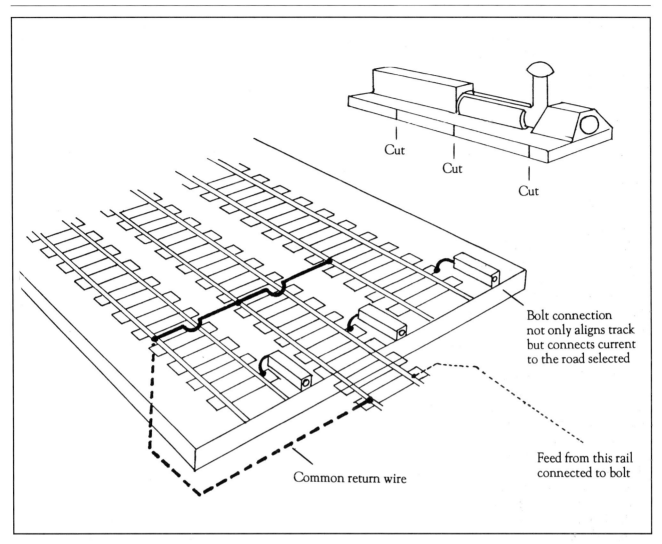

Cut

Cut

Cut

Bolt connection
not only aligns track
but connects current
to the road selected

Feed from this rail
connected to bolt

Common return wire

A method of locking and providing an electrical contact between the main baseboard and a sector plate or traverser. An ordinary metal door bolt, approximately 3 inches long, is cut down to form sockets. A single bolt should be sufficient for three tracks.

building (see pages 75-77). Similar considerations to the traverser apply to lining up siding to entrance/exit roads and to the provision of a suitable electrical supply.

One almost final point on fiddle yards - generally speaking, the length of the fiddle yard road will govern the length of the train that can be run. There needs to be some correlation between the length of the fiddle yard road and the platform. If you have a four-coach train in your platform and your fiddle yard will only take a two-coach train or vice versa, I would suggest that you've got your design wrong somewhere!

The final point on fiddle yards is to watch the transition from layout to fiddle yard. We tend to skimp on the fiddle yard and it is all too easy to end up with improperly aligned baseboards or trackwork. It is surprising how many layouts have trains that lurch across this great divide. I know - I've got one at the moment that needs some attention in this area!

Returning to the plan that evolved for the project layout, I am sure it won't have escaped your notice that there is little in the way of pointwork at the station throat. Here is another advantage of the use of traversers or sector plates - they can also double up as an off-stage set of points used to close a run-round loop or to change from up to down line or vice versa when shunting or carrying out similar train movements. In the case of the project layout that is exactly what the sector plate does.

Operational and visual interest

The plan as it has evolved provides a variety of facilities that will help operating and visual interest.

It was envisaged that the two main platform faces would be typical of the type of facility offered in the imagined location and for the role of the station, ie an important but secondary station serving a large town. The third platform drawn in the plan is really to provide for parcels and van traffic. It could, as drawn, also provide the base for a DMU or auto-train shuttle.

A crossover is provided at the platform end, in keeping with the prototype. I assumed at first that arrivals would normally enter platform 1 and normally also depart from that platform; I provided in the plan a loco spur which would enable a loco to back down on to carriages at platform 1. The train would then depart across the crossover on to the correct road out of the station throat to the fiddle yard. Similarly, a pilot loco could hook up to the carriages and shunt them either to an imaginary carriage siding off-stage or to other platforms, thus releasing the incoming loco for either the return journey or off-stage for turning and servicing (not, of course, necessary for a tank loco). I think, therefore, that the crossover adds to the layout and enables the loco spur to become a short storage road.

I have an aversion to seeing platforms filled totally with a train, something that gives the appearance of the train dominating the layout rather than the layout dominating the train. It is, I acknowledge, very difficult on a small layout, particularly where, as in this case, we're trying to squeeze the quart into the proverbial pint pot!

Avoiding filling the layout with a train does help the illusion we are trying to create, as does the treatment of scenics - of which more later. The platform will take a train of three 48-foot or 50-foot coaches and a medium-sized 4-6-0 with plenty of room to spare. Three 57-foot coaches will fit at a pinch, but British Railways Standard 64-foot stock is definitely out.

I had envisaged that the layout would principally be, at least as far as passenger services were concerned, operated by tank engines - for example, Ivatt 2-6-2s and Stanier, Fowler or Fairburn 2-6-4s for the 1960s London Midland Region. Thinking of a grimy Stanier tank and two or three odd coaches simmering in the platform brings back memories - ah, I can almost smell the scene and taste the grime! Earlier periods with 2-4-2 and 0-6-2 tank locos and shorter coaches would make the space problem easier.

Other regions and locations would reveal similar combinations, perhaps a large 'Prairie' or pannier

An Ivatt Class '2' 2-6-2 tank Loco in 7 mm scale, a key locomotive type for our project layout. The building of this loco and, indeed, all the locomotives and rolling-stock used on the layout are detailed in a later volume.

Above Short-wheelbase Fowler 0-6-0T dock tank No 47165, specially designed for sharp bends and tight spaces, so ideal for our kind of layout! *D. Hampson*

Below Crewe special tank No 3304 in LNWR days. Such a locomotive would be very appropriate on the Platt Lane layout as this class would have been used for freight, trip workings and shunting on LNWR and LMS Western Division lines from the 1890s to the late 1950s. There is a well-known photo of one that went through the buffers and crashed into houses in Crook Street, Bolton, close to Great Moor Street, our source! *Author's collection*

As mentioned before, don't forget the continental prototype. Here are two rare and unusual photographs of engines being turned by hand in the early 1900s. The tank locomotive in the upper picture is PLM 0-6-0T No 3640, while the tender loco is PLM 0-6-0 No 1282, presumably being turned on the same turntable. Because of the latter's small size, the tender has had to be uncoupled and turned separately. Note that the *mechanicien* stays on the footplate while others do the donkey work! *Author's collection*

tank on the Western Region, an 'H' or 'M7' Class on the Southern or a BR Standard Class '4' tank - the possibilities are considerable.

The freight facilities would be split between the coal drops adjacent to the passenger facilities and what I have termed Bollings Yard Sidings. Here, any number of facilities could be developed and I tend to favour somehow combining the fiddle yard screen with the development of say a factory, mill or warehouse wall.

Another possibility would be a row of terraced house backs, backyard wall, boundary wall and some form of sidings. The simple arrangement of a few sidings, railway yard offices, crane and ancillary huts would, I think, give a pleasing functional and wholly compatible facility, although some rather more ambitious ideas are suggested by the accompanying photographs of different goods installations. It could also provide a contrast in the use of materials - say, a stone boundary wall against the brick used in the construction around the station area.

The biggest problem at this end of the layout is to

A transfer freight of a handful of wagons hauled by an ex-L&Y 0-6-0 in 1960, typical of the type of train we might use on the project layout. There is plenty to note in this shot when it comes to scenic detailing for the layout in terms of ballast and weed growth. D. Hampson

provide a convincing screen to the entrance to the fiddle yards. The long frontage mentioned in the last paragraph presents an opportunity for development, the screen running from front to back a liability.

It is quite possible to avoid this difficult screen and to disguise the entrance by a strategically placed building or other structure which would naturally take the eye away from the hole the trains came in and out of. The logical structure to do that would be a large signal cabin. Space is very limited between the tracks, so perhaps one with a narrow base and an over-hanging top - a bit like medieval houses - or perhaps, more convincingly, a signal cabin on stilts over the track.

The simplest arrangement would be an overbridge running from front to back, perhaps at an angle, carrying another railway line. The rising contours from the roadway to the Bollings Yard level would give credibility to the need for this level. Such a bridge would take the eye away from the station, making it feel less cluttered but, conversely, may be too strong an attraction.

The road bridge and the concave vista of the raised trackbed level naturally draw the eye to that part of the layout so, if anything, visual balance needs to be created at the opposite end of the layout, the platform concourse end. I'll leave it until much later to let you know what I actually did.

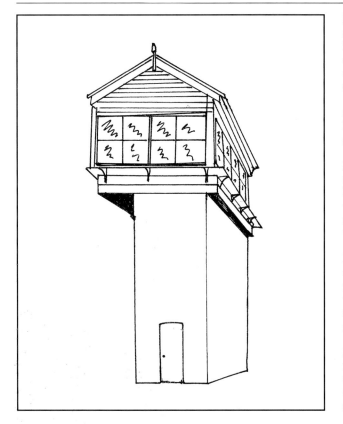

Left Hiding the entrance to a fiddle yard can be a problem. I don't particularly like the standard answer of a tunnel or overbridge, although I contemplated both for Platt Lane. A removable Victorian park over the fiddle yard, which was tunnelled under by the railway, and a bridge carrying another railway over the station throat were possibilities.

However, I opted to hide the entrance with a building. A signal box seemed ideal, either a tall LNWR-type on a narrow base between the tracks or one above the track on timber or steel supports. An example of the first is shown.

Below Or try modelling this! A fine photograph of a tar distillery as a source for an industrial model, providing enough to keep you occupied for a long time with all the pipework. It would also be a good excuse for running some special wagons, not least the many tank wagons now on offer, very colourful in their early liveries.

Right A timber yard adjacent to the Manchester Ship Canal, complete with derrick and a host of equipment. The rail-mounted crane in the centre is interesting, showing a proper use for a type of vehicle often modelled but seldom placed in a realistic setting. Note also the interlacing of sleepers on the point in the bottom right of the picture. *John Harmon*

Below right The following three photographs were taken at different locations within Manchester Docks showing a variety of features that could be incorporated on a model dock layout. The first shows a rail-mounted crane, and wagons being loaded from a ship. Note the railway track set into the stone sets. *John Harmon*

Above left The second view shows a train of company-owned wagons being loaded with timber. These wagons would be ex-railway company and, indeed, the second vehicle from the left is of Midland Railway origin. *John Harmon*

Left The third photograph shows tea being off-loaded from a ship to the canal boat adjacent to the quayside railway. The large dockside cranes could be moved up and down the dockside on the railway line for unloading and loading of cargoes. *John Harmon*

Above A connection between Bolton and the Manchester Ship Canal, and a further idea for a small industrial installation: a wagon repair works and Ship Canal wagons. *D. Hampson*

Picking up on the theme of the station building end of the layout, Great Moor Street for some reason had what was, so far as I know, a one-off Italianate-style building of some note, alas no longer with us. It seems strange that such a building should have been tucked away at a secondary terminus surrounded by other buildings and structures drab even by railway standards.

This building was across the platform ends, the first floor at rail level, the ground floor at road level. As you can see from the photographs (page 29), there wasn't much in the way of ornate canopies or other paraphernalia outside. I have, however, always quite fancied the idea of modelling a simplified station concourse with taxi ranks, etc, and here was an opportunity. The provision of such a facility takes up a great deal of space, proportionately; I therefore developed an idea to make this a separate bolt-on board, no more than 28 inches or 2 feet wide but spanning the full width of the layout. The thoughts behind this were first that this scene was not essential to the operation of the layout and therefore if space were at a premium at home it could be left off, but added on when the layout was used at a show. Second, on its own it could be quite an attractive model which could, when not in use on the layout, be displayed in the house or railway room in its own right. It could be protected by a Perspex or similar cover which could be arranged to clip at the bottom over the baseboard.

The station canopy provision needed careful consideration. Attractive though they may be, canopies are devils to build - and they have to look right as they are so conspicuous. There is also the problem that if you are not careful you hide the train, a very real problem with overall roofs and a good reason why they are seldom modelled. The type I favour is shown in the illustrations overleaf, and variations on this theme were used by virtually all the British railway companies. An easy way out would have been to set the layout post-1962 when in all probability all that would be left would be the framework, or even just the holes where the upright supports once were!

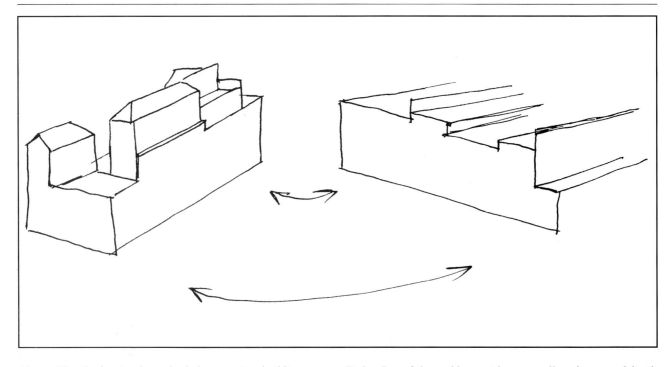

Above Sketch showing how the bolt-on station building concourse is arranged. Great care is needed to ensure a match between the detail on the two boards, particularly the platform ends, walls and street levels, etc.

Other railway structures would be limited to the signal box and various huts and small goods offices. These would be in the style of the company which built the line. There are plenty of drawings and photographs of the many railway structures in the model railway press and in books.

The Bollings Yard area could be a site for a railway-owned goods warehouse or medium-sized goods shed and here again, reference to the railway and model press will no doubt provide a suitable reference base.

Atmosphere and ambience

To create the right ambience and impression with a model railway you have to adopt a consistent approach, not only to the overall standard of modelling but also to the things you put on it. Unless you are going to model an imaginary preserved railway, you won't create the image if you are inconsistent with your fixtures and fittings. It is an anathema to have a GWR 'Castle' with two Southern Maunsell coaches running into a station that is clearly, say, Caledonian in origin but with LNWR signals. That is the quickest way to kill any atmosphere.

Generally speaking, a railway built in the 1880s would keep its original buildings, water columns, yard cranes, buffer stops, etc, unless it was the subject of a major rebuild. Even then, only what was necessary

Right One of the problems with an overall roof on a model railway is the need to see the locos, the stock and the modelling, not hide them. Canopies and high walls to protect passengers and goods are not conducive to this. Clearly, therefore, an alternative is needed and these sketches show some possibilities.

was removed or replaced. Subsequent modernisation programmes may have perhaps changed the signalling to upper quadrant semaphores, and an odd building or fitting may have been replaced - but only when absolutely necessary.

Referring back to the little Kent & East Sussex Railway layout that I owned, while the layout represented an imaginary though unlikely possibility, it had buildings, structures and scenery that were all compatible and consistent with the K&ESR and its environs. To have brought into the model other types from different sources would have immediately destroyed the intended image. Similarly, the locomotives were exclusively all types which ran on the K&ESR during the 1950s and '60s, the period during which the model was set. The coaching stock consisted of a single Birdcage brake of the type (and even the same number) as the one that provided the mainstay of passenger services for a time. Goods trains were short and made up largely of open wagons, which photographs of the period show to have been typical. No room for colourful and fancy tank wagons here which would spoil the picture I was trying to create! The same sort of discipline or restrictions would have applied if I had chosen to model the Mid-Suffolk Light Railway, the Loch Tay branch or indeed any line, route or location.

So far as the project layout is concerned, if, let's say,

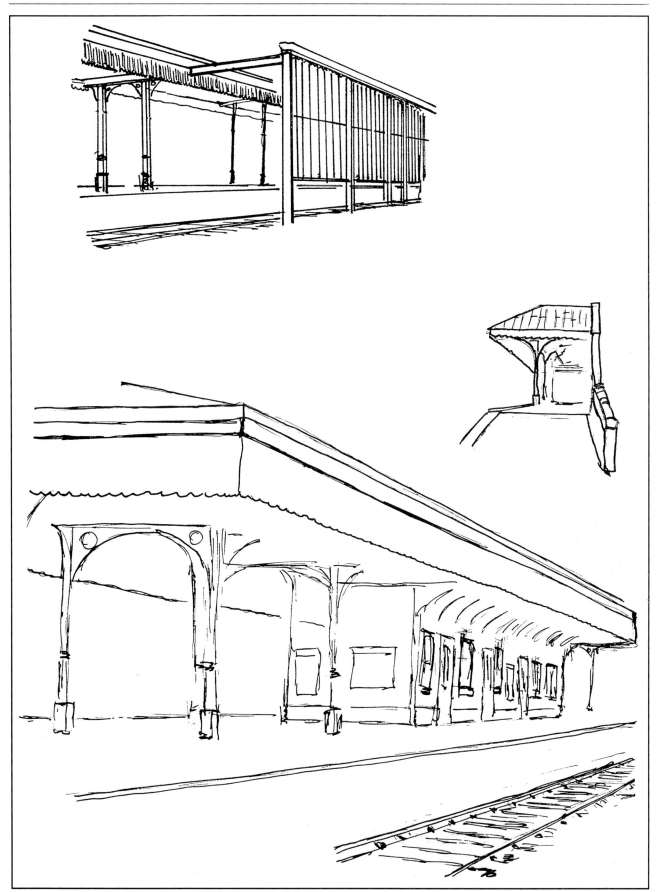

The partly demolished station site outside Great Moor Street, the Platt Lane inspiration, full of detail for the modeller. The shops that butted on to the coal drop roads adjacent to the platforms were there until recently, betraying no trace of their railway heritage. *D. Hampson*

its origins lay with the London & North Western Railway, it would clearly have many features that would betray that origin. Indeed, the photographs of the prototype show this.

Up to the Grouping of 1923, the fixtures and fittings would show a pure LNWR origin. If you wanted to be pedantic it is possible that some would show the characteristics of a particular period in that company's development - certainly with even the smaller railway companies, particular lines and routes showed a different style. This may have been due to local conditions, the use of local materials, or perhaps origins in a specific small local railway-building enterprise which became absorbed into a bigger system later. Signalling and signal boxes might be an example where perhaps these changes occurred; independent railway signalling contractors such as Saxby & Farmer may have been contracted to build signal boxes which would exhibit their style rather than the one that became associated later as being a characteristic of the railway company itself. Liveries of buildings, lettering and signing styles were also prone to change and development.

In our source, the signals, signal box, yard crane, water crane, signs and notices were to change little until well into the LMS period. By the late 1920s the railway signs and notices would have changed, and when the LMS 'totem' pattern was developed in the late 1930s may have changed again. The British Railways period would see the spread of the ubiquitous regionally coloured enamel signs. Liveries of buildings would change too, but the process of change, beyond stamping the name of the new owner prominently on things, was a slow one. Many parts of the former LMS never received, for example, the new LMS standard

totem signs of the 1930s, and doubtless some structures not a coat of paint between the LNWR and BR!

We know from the photographs that the signals were changed to the upper quadrant LMS/BR standard types at some period, probably in the 1950s. The LNWR water cranes would stay to the end as would also, unless from absolute necessity, the crane in Bollings Yard. The cast iron notices might also survive with only the name of the original owner painted out.

It is possible that the signal box could have been replaced either by the LMS or by a BR type, or perhaps a hybrid of a new wooden top on the original brick base - so there is some scope to break away from the original mould. Perhaps in the late '50s the gas lamps might have been replaced by electric lamps on concrete posts, and the platform canopy might similarly have been replaced with a concrete and steel cantilever affair. Certainly if the station had survived into the 1970s a bus shelter and single platform with colour light signalling amongst urban decay so beloved of social commentators could be a serious modelling challenge.

Enamel advertising signs were very long lived, products and advertising campaigns of the pre-Second World War era being very enduring. Virtually all the enamel advertising signs reproduced in model form would be suitable for the first 40 years of this century.

While I have concentrated on a scenario of possible changes for the project layout in its Lancashire setting, similar changes would have occurred across the country and, as I have said before, you can change and adapt to suit your preferred location and circumstances - it is only an example that I am offering. In particular I have tried to outline the areas that need consideration before building work commences on

Photographs such as this are quite rare - no three-quarter front view of a loco! They are, however, invaluable in providing detail for the modeller. Mention is made in the text of the often slow pace of change for many railway features. Here in 1963 at Moorside & Wardley there is still a fine collection of Lancashire & Yorkshire signals, an L&Y signal box, and what looks like an LMS platelayers hut. The signal in the foreground has the white diamond fixed to the post indicating a track circuit. Note also the ballast and the appearance of the ground between the tracks; that on the right has obviously been recently ballasted. *D. Hampson*

the layout, and in some cases before the track plan is finalised. The railway buildings and structures have a centre-stage role, and are just as important as the locomotives and rolling-stock in creating an impression of the railway we are trying to represent in model form. Just as important, however, is the rest of the stage, the setting for the production.

The impression the modeller is trying to create extends beyond the railway boundaries to the setting in which it is located. Even if it is only a railway embankment, a station master's house or terraced house backs, the setting becomes an integral part of the scene and therefore it is necessary to maintain that overall standard of consistency to these items.

L&Y ground signals, bullhead rail but new brickwork - Blackburn Station, July 1963. *D. Hampson*

A railway embankment may seem just like a railway embankment anywhere, but the type of ground will determine its angle and, where it is shown through the undergrowth, not only its colour but also its texture - a cutting through chalky downland will look different from one through Pennine granite.

The non-railway buildings will also look different. Public houses, for example, vary quite considerably, often having characteristics peculiar to the locality. A pub is not just a pub but a building which should reflect the pubs in the area you are representing in your model railway. A half-timbered pub of the type so beloved of plastic kit manufacturers would look just as incongruous at the end of a row of terraced back-to-backs as would the 'King' on the Loch Tay branch.

Terraced houses, whether you are modelling their fronts or backs, are just as much a product of their function and locality as the railway. Local materials, local fads and fashions of the time they were built, and the status of their intended occupants all give them a flavour that should subtly enhance the total scene you are creating and help make sure that the location of your model is clear for all to see.

These houses, ubiquitous and humble, varied tremendously in detail and from area to area within a town. The accompanying photographs show two types of terraced housing of similar status and age in comparable towns less than a dozen miles apart. The one thing monochrome photographs do not show is the subtle differences in the colour between the two, which also gives a clear indication of their location; many an otherwise excellent layout is spoiled because of this. A terrace in one part of Salford is as distinguishable from one in Bolton as to one in Exeter. Never mind the authenticity of the street lamps, tramcars and railway buildings - the authenticity of the local housing is equally important.

The area on the layout behind the platform wall will provide a street scene with a large factory frontage and terraced houses based on those which surrounded Great Moor Street. The area at the fiddle yard side of the bridge could be developed in a number of ways. Backs of terraced houses with detailed backyards, a small factory, even a combination of the two, could be located there.

On the same side of the bridge but on the frontage of the layout I envisage a cinema fronting the road, not a grand Odeon but rather a 'flea-pit' from the silent era with its wrought iron and glass canopy, plain walls and grand entrance. Definitely a budget flea-pit, perhaps reminiscent of the one in the 1957 Peter Sellers film *The Smallest Show on Earth* where

A typical industrial townscape, which could suit almost any period between 1910 and 1970 except for the car and dereliction. Compare these terraced houses with the next picture.
D. Hampson

A terraced house in Ordsall, Salford, before the last war. This shows a different style entirely from the previous picture. For all the ornate mouldings above windows and doors, this was a house in a poor district serving the docks. *John Harman*

the projector and screen shook every time a train passed over the adjacent railway viaduct!

Behind that, as we approach Bollings Yard, I envisage a short row of terraced houses, fronts facing the railway with detailed backyards and allotments filling the surrounding space. This site is on an incline down from yard to roadway level and I think lends itself to these usages.

By now I hope that I have given you food for thought in planning and designing your model railway, adopting an approach to the process that is holistic in concept rather than just an emphasis on the railway itself. I think perhaps the approach could be likened to that of the impressionist painter - don't concentrate too much on the detail, but the whole picture. It is the *impression* you are trying to create.

The scenic ideas discussed for the project layout have necessitated quite significant changes in levels of the scene surrounding the railway, and this needs to be considered in arranging for a suitable baseboard and presentation of the layout. Careful planning clarifies what's required from the baseboard and can save a lot of work and aggravation later by providing for this at the design and construction stage. Which leads nicely on to the next section - baseboards. There will be a lot more on buildings and scenics later.

Typical of a street scene incorporating shops and houses in a terraced row. The picture is believed to be *circa* 1930 and was obviously taken on a snowy winter's day.

3.
BUILDING THE BASEBOARD

Planning ahead

The foundation of any model railway is the baseboard, and a well-made baseboard is an essential prerequisite for a successful model.

Model railway baseboards come in a number of styles, and methods of construction vary; the most common variations are illustrated in this chapter. The design of a model railway, its shape, the type of landscape in which it is set and the use to which the model will ultimately be subjected will all influence the design and construction method of the baseboards.

This latter aspect, that of the ultimate use of the layout, should not be lightly assumed as it is often necessary, for unforeseen reasons, to move a model railway. Accordingly, it is sensible, even with a permanently located layout, to consider and plan for a possible need to move it, at least in part. Similarly, if a layout is located in a loft or a shed, the baseboards themselves must be brought into the areas in which they will be located. It is therefore impossible to take, for example, a 4 ft x 2 ft 6 in baseboard through a 2 ft x 1 ft 6 in loft hatch. Regardless of loft hatch access, it would not be wise, for obvious reasons, to consider baseboards bigger than will fit through an average door opening (6 ft 6 in x 2 ft 6 in).

The size of baseboards is also related to the layout design, not only in the obvious sense of needing a base on which to locate the necessary trackwork and scenery, but also in the sense of the location of the baseboard joints to avoid difficulties with trackwork or in hiding joints with scenic work or buildings. While it is perfectly possible to have baseboard joints occurring beneath pointwork by the simple expedient of cutting the track, it is something best avoided! Baseboard joints, particularly in portable layouts, can be difficult to mask successfully at the best of times, so it is wise to try to avoid creating obvious difficulties by, for example, having baseboard joints which of necessi-

ty go through the middle of a pond or a large and prominent building. I know buildings can be made to be demountable, but this brings its own problems and solutions which are more properly discussed later.

You may have noticed that I talk of avoiding the baseboard joints cutting through scenic features, rather than constructing the scenic features to avoid the baseboard joints. The construction of a model railway is a compromise, but the approach I am suggesting, with careful planning and preparation, will show up the constraints on the location of baseboard joints, which are a significant element to be borne in mind when planning the layout.

It becomes clear that the more thought that is given to constructing a model railway, the easier it is to construct. The design process takes the concept of the layout we want to build, the 'vision' if you like, and develops it through a variety of processes to achieve the end result. The design of the baseboards and their method of construction is clearly an integral part of the early design process.

The baseboards for the Platt Lane project designed for this book are built principally from 9 mm plywood, with a very small amount of softwood for corner bracing and thinner plywood for specific features. The illustrations and photographs will show fairly clearly the construction but, to understand fully their evolution, they need to be looked at in conjunction with the plans and sketches of the layout design. For what it's worth, I set out a cutting list of the materials used for the baseboards for the basic 16 ft x 3 ft of the Plant Lane project. Cost is seldom mentioned in respect of the building of a model railway, but the timber in the cutting list at mid-1992 prices was £118.

In terms of the supply of such materials, if at all possible get them from a timber merchant and not a DIY superstore, which is generally much more expensive. While I appreciate that cutting sheet material at home can be difficult, and some DIY stores offer a

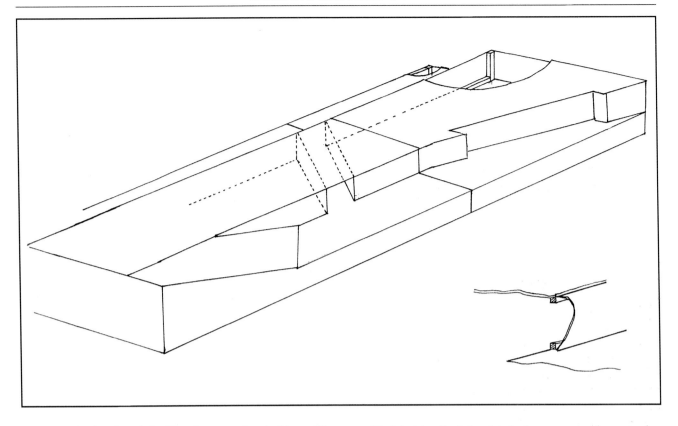

The two main baseboards for Platt Lane are sketched here. The cut-aways show the joints using three-quarter-inch square softwood, screwed and glued. The sides and ends on each baseboard are cut in one piece to the required profile. The upright pieces are supported on softwood brackets shown in the inset, as are the bridge underspans and the incline from road level to trackbed height. Each level is built up separately; note that apart from the end pieces and bridge underway, no single piece of plywood deck crosses the complete width of the layout. The sides, ends and trackbed are made from 9 mm plywood, the uprights and retaining walls, etc, are 4 mm plywood. All joints are screwed and glued where softwood is used.

cutting service, it has been my experience that a timber merchant will also supply sheet materials from a cutting list, usually at no extra cost. The quality of timber tends to be better and there is a choice of grade and finish.

Earlier, the construction of mock-ups of proposed layouts was mentioned, and I would again reiterate that it is a good idea to do this at baseboard design stage, particularly if there are complicated scenics or if the basic scenic structure and shapes are to be incorporated into the trackwork as in the case of the Platt Lane project. Care taken in the design, planning and construction of a model railway, and particularly in respect of the baseboards, will pay dividends later.

A final general point relates to the movement of portable layouts. There is clearly a limit on the amount and size of materials one man can safely lift, let alone carry, especially when negotiating awkward stairways and entrances. If a model is being built as a team effort or as a model railway club exhibition layout, where it can reasonably be expected that several people will be available to carry baseboards, different considerations regarding size can be contemplated.

Construction methods

The tried and tested methods all revolve round timber. The traditional wisdom for baseboards goes along the lines of a solid-top affair, usually chipboard or similar material, with a softwood frame, usually of 2 in x 1 in. For obvious reasons, this type of base and its similar variations only allows a limited amount of development below baseboard surface level, for example for rivers, canals, etc. There is more scope for above surface development, but even a hollow framework for scenic work above this solid surface is quite wasteful of surface material.

The maximum amount of flexibility comes with a completely open frame system. The principle of the 'L'-girder system was developed in the USA. It uses softwood risers to support trackbeds, etc, attached to a network of softwood longitudinal and cross pieces. Track and road beds are cut from plywood only to the extent necessary - the Americans call this the 'cookie cutter' method. The system gives maximum flexibility in scenic provision, can be adapted to suit virtually any location irrespec-

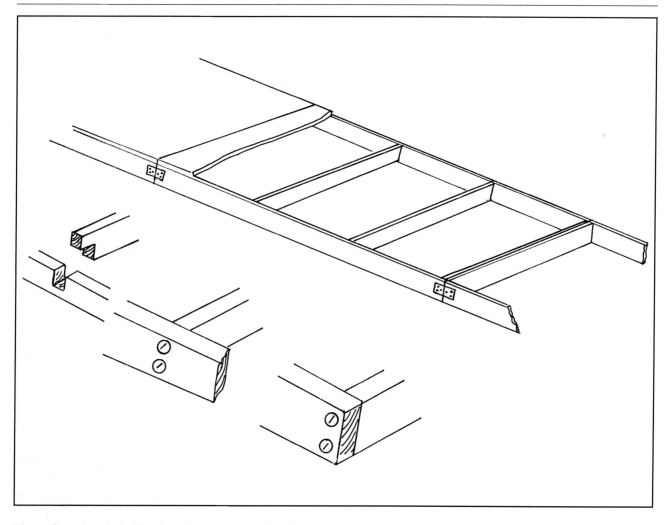

The traditional method of baseboard construction - ½-inch chipboard on a 2 in x 1 in softwood frame. The baseboards are shown joined by the use of hinges with removable pins. When the hinge has been screwed between the boards, the head of the hinge pin is filed off and replaced by a longer piece of wire of the same diameter, the end of which is bent to stop it falling through.

Also shown are two types of simple joints between the intermediate cross members and the side pieces, and a third joint for the corner of the baseboard, which should be screwed and glued for extra rigidity.

tive of size or shape, and can, with a little adaptation, even be drawn on for the traditional British sectional portable layout.

The actual detail of plywood-based constructions varies and the two basic methods, each incidentally showing applications to different designs to which in my mind they are best suited, are illustrated.

Current thinking on baseboards tends to revolve around variations in the use of plywood, and in some respects is a compromise between the two systems just mentioned. Plywood has the advantages of being adaptable to a variety of shapes and contours of the terrain within which the railway runs, and above all is relatively light in weight, yet strong. Lightness is an important consideration if a sectional layout is to be moved regularly, say for exhibition use, and of course the inherent strength of this type of construction if carried out correctly

is also a major benefit. Anyone who has carried traditional chipboard and softwood baseboards up flights of narrow stairs into exhibition halls, let alone dropped the corner of a baseboard on his foot, will testify to the need for concern in these two aspects!

To get the best advantage from the 'L'-girder type of construction, a fair degree of thought and planning is required, which again emphasises the need to plan ahead and have a clear idea of the finished layout in mind. Scenery bases, including those for buildings and other structures, and contours can all then become an integral part of the baseboard with a potential for ease of achieving the ultimate scene required and the effectiveness of the finished model railway.

Above all, we cease to be members of the Flat Earth Society and can begin to breathe a little life and credibility into our creation. Here I am harking back to my early comments that in my experience the most effective and satisfying model railways are those that create an overall impression - providing of course that they run satisfactorily, and of course a good, well-designed baseboard is also an essential prerequisite for that.

Plywood of 9 mm thickness is my minimum for the structural parts of the baseboards, but other ancillary sections for, say, retaining walls and roadways can be of thinner material. It does not matter, of course, if the main thickness is used for these ancillary bits as well if there is some spare on the sheet, but the thinner material is suggested if additional material is needed as it is a little cheaper!

Choose the best quality you can buy - this will be well rewarded with easier working, greater stability and certainly a better appearance if outside edges

are to be varnished to improve presentation.

Main cross members can have holes or other bits removed from their centres to reduce weight without substantially weakening the structure. The method of fixing together baseboard materials is shown in the following diagrams and photographs and relies on the use of softwood joint pieces and glue-and-pin assembly.

We have already discussed the different types of fiddle yard, and the chosen system will need to be taken into account in the design of the baseboards. The easiest type to deal with is a fan of sidings stemming from points or pointwork or sidings laid out in parallel loops. Clearly these require appropriate widths and lengths of the material chosen for the main trackbed. The basic problem with this type of fiddle yard is that it is not very economical on space.

Earlier we discussed various alternatives such as sector plates, turntables and traversers. These solutions require specific approaches to baseboard construction, the principles of which are again shown in the accompanying sketches and photographs.

A multi-level baseboard, but still using the traditional solid surface on a softwood frame. Only a limited amount of sub-baseboard development is possible, as the river bridge cut out here, with its own sub-base. Note the provision of a sector plate fiddle yard on the left.

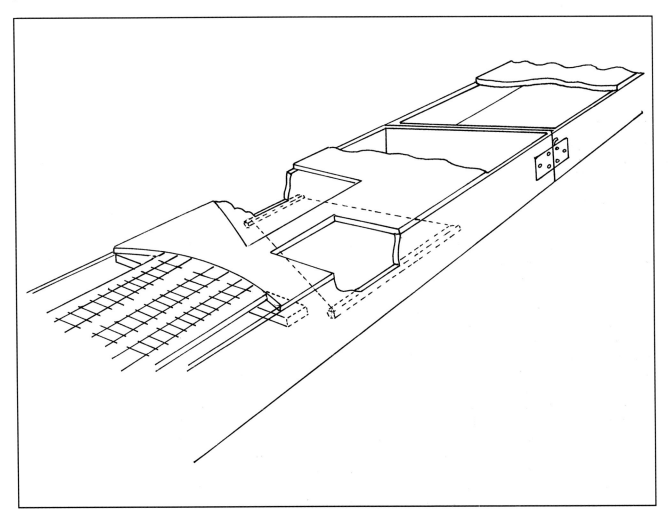

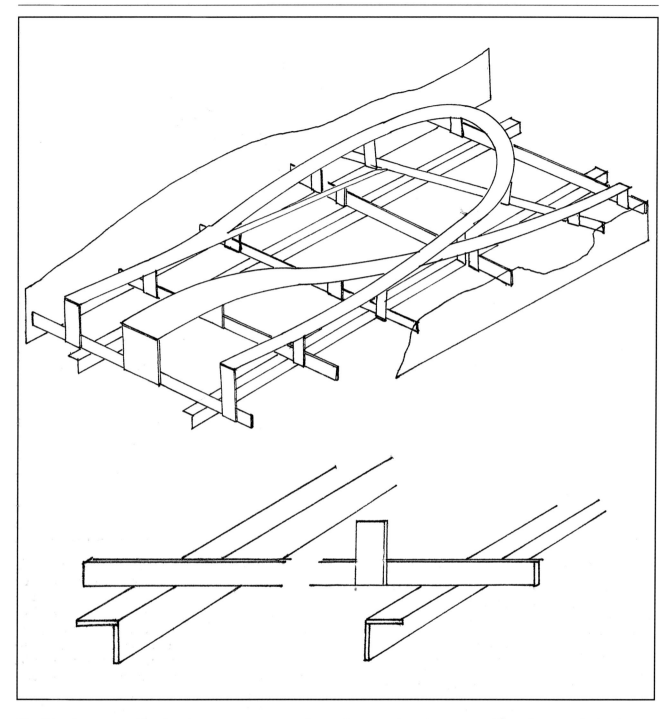

The 'L'-girder system of baseboard construction originated in the USA and gives total flexibility in locating track, stations and scenic features, and above all allows almost unlimited potential to develop features above and below the average track level.

The system is based on two longitudinal sections fabricated in an 'L' shape from lengths of softwood or plywood screwed and glued together (see lower sketch). Cross members are then screwed between the girders, to which risers are screwed to support plywood trackbeds. The cross members can be attached between the 'L'-girders at any angle to allow the development of particular features such as a wide gorge or river bed.

The edges of the layout are fitted with profile boards at the front, matching the scenic profile; at the rear and sides this profile board could be cut with a high straight edge to form a backscene with sky, etc.

In the USA it is normal for the 'L'-girders to rest on substantial timber legs bolted inside and underneath the girders, but this could be replaced by the more traditional British 2 in x 1 in folding legs, provided of course that they can be folded under without hitting a low scenic feature! An alternative would be to arrange for them to be detachable, bolted on and off.

This system really comes into its own with the multi-level track and complex changes of gradient associated with American models.

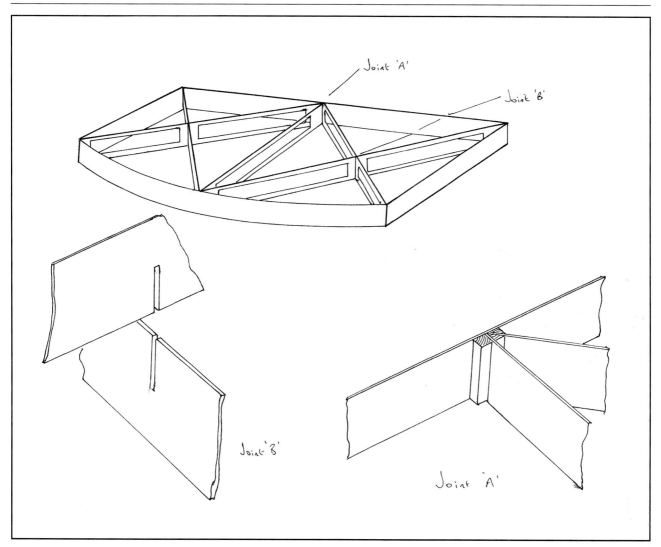

Plywood construction can provide a light yet strong baseboard. This sketch shows a more advanced type of construction that can be adapted to suit irregular shapes such as a curved edge.

The weight can be further reduced by the removal of the centre of the plywood cross members with little effect on strength. An alternative to the slots shown would be a series of large holes cut with a hole-cutter. Leave at least three-quarters of an inch above and below the holes, and a good 2 inches at either end. The two principal joints at A and B are shown in close-up, and should be glued and, in the case of A, pinned for strength.

Joining, moving and supporting

Having chosen the type of baseboard construction for your layout and its design materials, etc, you may be forgiven for thinking that your problems are over. However, there are two further aspects that need consideration, preferably before a sod, so to speak, is cut! These are the method of joining baseboards, and the method of supporting them.

With regard to the former, there are a number of approaches. The one currently in vogue, and with good reason, is the use of pattern-makers' dowels. This system allows the accurate alignment of baseboards across irregular joints such as those used in what could hardly be better described as the jigsaw design of baseboard. ('Jigsaw' baseboards are those that are irregular in shape rather then the more familiar rectangular design; their shape will have been designed to accommodate a specific scenic feature.) However, while the dowels provide accurate alignment they do not hold the baseboards together! Accordingly, some method of fixing is necessary. The simplest expedient of all is to bolt the baseboards together using coach bolts.

The 'jigsaw' design allows for the baseboards to be designed in virtually any shape and configuration to help with the scenic or visual side of the model railway's development. It has the advantage of allowing baseboards to be designed and built to suit their visual purpose and minimise difficulties which regular straight baseboards can emphasise, for example in

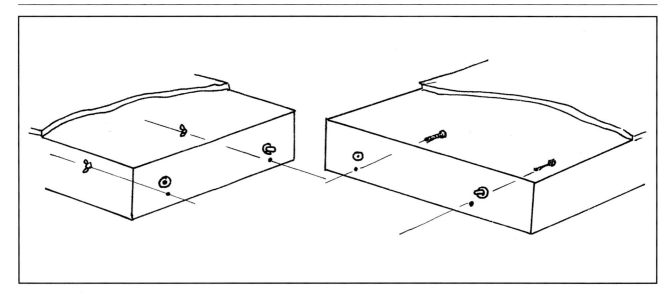

Above Pattern-makers' dowels are used to locate accurately adjoining baseboards, and beneath them are holes drilled through both end pieces to enable them to be held together with bolts and wing-nuts. Washers should be placed on the bolts on both sides of the joint to protect the wooden ends.

Below An end-on view of the second baseboard looking towards the fiddle yard and the third baseboard. The end piece, cut from a single sheet, is contoured to the shape required. I would usually cut a pair of ends together, the second piece forming the corresponding end of the adjoining board. It also makes it easy to clamp the two pieces together, ensuring perfect alignment, and to drill through both at the same time to provide the location for the pattern-makers' dowels. The male and female dowels, one on each side, are clearly visible, together with the corresponding bolt holes for the bolts which will hold the baseboards together.

disguising joints across platforms, rivers, etc.

The disadvantages, for there are indeed disadvantages to everything in the exercise of the art of compromise which is the essence of railway modelling, are principally related to the complexities of erecting and dismantling layouts built this way, together with their support, carriage and transport. There is also once again the matter of the need for a carefully developed and planned approach to the design of the model and a clear idea from the outset of its intended visual appearance - none of which in my mind can be a bad thing.

Clearly, any reasonably sized layout complex enough to be interesting operationally and visually is going to require a number of baseboards. In the 'jigsaw' these are by definition not of uniform shape and the problems of carriage and storage of these components, while not insurmountable, need careful consideration and planning.

Some form of box or carrying frame designed to protect baseboards in these circumstances, while also making them easy to store, handle and transport, becomes essential. A simple arrangement is outlined in the accompanying sketch.

If in designing a baseboard the key elements must include a rigid, stable but lightweight structure, then this must also be applied to the support system. The key point in designing and building legs, trestles or other variations of support, however, is that layouts, properly constructed and even with stock and control equipment, are not very heavy. Accordingly, they do not need to be supported on 4 in x 3 in legs!

An exterior close-up of the 'female' part of a pattern-makers' dowel used to align the baseboards.

An interior view of the housing for the dowel. The wooden block shown screwed and glued in place on the baseboard end piece strengthens the area around the dowel which, when seated into the end piece properly, almost comes through the plywood. The hole beneath the block is for a bolt and wing-nut which hold the baseboards together.

Support for model railway layouts comes in several different arrangements and should have the aims of giving stability to the assembled layout and being easy to construct, accommodate and transport. I have assumed again in the foregoing remarks that the layout will be portable. One of the accompanying sketches shows the arrangements I have come to favour for a site which can more or less be regarded as the permanent home for the layout. This method has the advantage of assuming a level framework on which baseboards can be rested. I have indeed built exhibition layouts supported on this type of construction for home use and while being worked on, but have kept separate trestles or indeed folding legs housed within the baseboards themselves and folded down for use when the layout is taken out and erected elsewhere, such as at exhibitions. A variety of arrangements for supporting the layout are shown in the other sketches.

A method of transporting a simple shelf-type layout by inverting one baseboard over the other to form a box which is held rigid by plywood pieces bolted to both baseboard ends - use the holes that join the baseboards together where practicable. These ends also protect the contents from damage during transit.

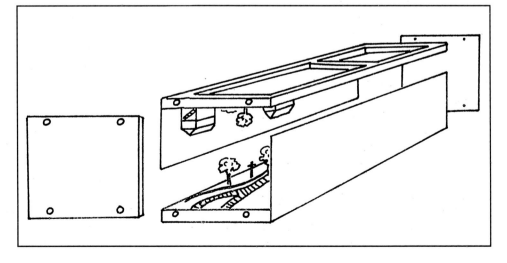

I tend to come back to the fold-under legs, self-contained within the baseboards themselves when folded up and not in use. This system can be quite economical, as only one baseboard need have two pairs of legs, one at each end; the remainder need only one set of legs as they are in turn bolted on to each other, the unsupported end being attached to the board with two sets of legs.

Trestles also provide an easy system to set up - simply set out the trestles and place the baseboards on top. I have always been concerned to ensure that the baseboards can't slide or gradually work their way off the trestles, and this can be achieved in a number of ways. The easiest to arrange is a locating slit in the trestles in which the baseboard frame can sit.

Concern about the stability of narrow baseboards, particularly if arranged in a straight line and at a reasonable height, has led to the adoption of some legs of substantial timber to try and lower the centre of gravity and make the layout more difficult to accidentally push over. Some years ago, more now in fact than I would care to acknowledge, I had a 16 ft x 1 ft straight layout with ordinary paste-table-type legs of 1 in x 1 in softwood which was quite stable. However, in an endeavour to increase stability still further, the legs were splayed out an inch or two at the bottom by the simple expedient of a fold-down crosspiece wider than the width between the legs at the tops, thus making the base wider than the top! This is shown in the sketch on page 67. However, I express a word of caution - while it worked in this less than careful case of youthful exuberance over science, it is obvious that there is no guarantee of the splay being equal on either leg, and thus it would seem quite possible to have a perfectly stable layout but one which slopes across its width. Having not had the call to build a long narrow layout since, I cannot say that this crude but effective device would always work without giving further problems.

I have recently been quite interested in the use of plywood for folding legs; this has been used to great effect on a local club's N gauge layout. The principle of the folding leg is merely applied to 6 mm plywood which has had a great deal of its centre removed to reduce both the overall weight of the layout and also strain on the hinges. It seems to me that this type of leg support could have considerable possibilities, and a couple of alternatives are sketched. Careful planning could, I believe, lead to the legs made in this manner folding flush with the underside of the baseboard and perhaps helping to protect the wiring, point and single mechanisms, etc, that lie beneath the baseboard surface.

A level surface and accurately aligned baseboards are essential both for good running and visual appearance, but beware - floors are seldom level, despite their appearance! There have been some very sophisticated methods used to overcome this problem, but more basic approaches should not be discounted before considerable time and effort is expended. First, scrap hardboard, plywood, etc, should always be carried with portable layouts to pack legs to achieve a level, well-aligned layout. Second, it is quite a simple matter to arrange adjustable feet, as shown in the accompanying sketch. Being by nature one who avoids unnecessary effort, I have always tended to rely on the former crude but effective insurance. Don't, however, assume that it is only exhibition halls that have unlevel floors. Experience in the last two houses in which I have lived, even though there was no subsidence, showed them to have less than level floors, hence the development of the framework bolted to the wall and mentioned earlier.

I'll conclude this chapter with a series of detail photographs of the construction of the project layout, showing some of the techniques employed.

Sketches of various methods of baseboard support.

1 Baseboard A has two sets of fold-down legs which, when braced, form a stable base to which other baseboards can be attached. Each subsequent board, as with B, requires only a single pair of legs, each of which requires bracing.

2 Simple softwood braced legs as used on the project layout.

3 An arrangement with folding trestles carried separately from the layout. They will require stays to hold them open.

4 Off-cuts of plywood of suitable size and thickness can be cut to make fold-under legs. Being thin, plywood enables one set of legs to be folded under the other; if fastened in place, this arrangement would help to protect the underside of the baseboard with its delicate point and signal control mechanisms and wiring. The cut-outs reduce weight.

5 Slot-in baseboard legs held in place with a nut and bolt and located between softwood formers (see also the photograph overleaf). Some form of bracing is recommended to strengthen the support; a strip of sheet material or stripwood fastened to the leg and the underside of the baseboard, as shown in sketch 1, would suffice.

6 A simple arrangement for adjustable feet. A bolt, of a minimum diameter of a quarter of an inch, is screwed through an angle piece, drilled and tapped to accommodate it and attached to the bottom of the leg; a captive nut could also be fixed underneath.

7 An arrangement for a simple frame screwed to a wall, with legs and frame of 2 in x 1 in timber, or less for the legs. Simple hammer-and-nails joinery would suffice, but ensure that it is level. It could be arranged to fold or be removable if the room, a garage for example, is used for another purpose.

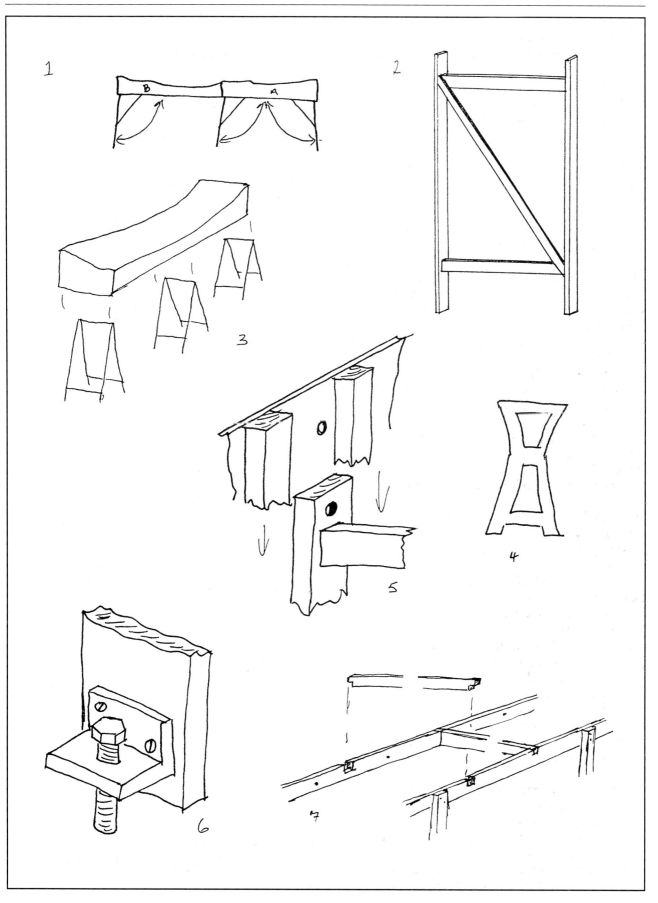

Left Traditional 2 in x 1 in legs and baseboard frames screwed and glued as shown in the sketches, awaiting the top surface.

Above Softwood fold-up legs have been used to support the project layout, and the arrangement for housing the legs is shown here, turned on its side. A simple box of ply and softwood is pinned and glued to the sideframes and drilled to accommodate a bolt on which the legs are mounted and pivoted. This bolt is put in from the outside, leaving the exterior face neat. Note that a second nut has been added to reduce the risk of losing one.

This is the housing for a fixed, ie non-folding, leg, again turned on its side. It is a simple arrangement using scraps of softwood and ply, screwed and glued in place. The housing must be a snug fit around the legs, which are held in place with the bolt and wing nut. It is surprising that such a housing need only be relatively short to hold the leg and restrict movement sufficiently to give stability to the baseboard - 2 to 3 inches is enough.

Right Basic carpentry is all that is necessary to build baseboards - it is the care and accuracy that goes into the construction which counts. Here, a basic joint in the upper cross piece of the leg is shown. The legs and cross piece are softwood, while the diagonal cross piece is left-over ply screwed and glued to the softwood.

Right and below The soft-wood leg and cross piece are joined by a mortice and tenon joint. Below we see how the tenon has a small wedge driven into the end grain to hold it firmly in place. The joint is also glued.

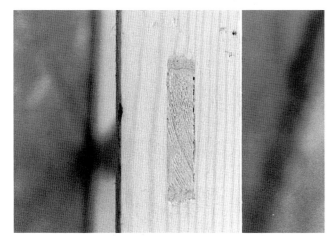

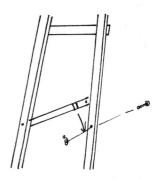

Right The famous splayed leg technique! The legs are of normal softwood construction, but a lower rail, longer than the upper one, pivots on a bolt on one leg and is fixed to the other by a bolt and wing-nut. Not very scientific, but it worked!

Materials list

Qty	
3	8 ft x 4 ft x 9mm plywood (best quality waterproof)
1	8 ft x 4 ft x 6mm ditto
1	6 ft x 15 in x $^3/_4$ in melamine-backed board
25 ft	$^3/_4$ in x $^3/_4$ in softwood
50 ft	2 in x 1 in softwood
20	M6 70mm bolts
16	M6 40mm bolts
30	M6 wing-nuts
1 box	$^3/_4$ in No 6 wood-screws
1 box	1 in No 6 wood-screws
1 box	$1^1/_4$ in No 8 wood-screws
6 sets	pattern-maker's dowels

These are the main purchases of timber, screws and iron-mongery used for the Platt Lane layout, and illustrate the kind of quantity required for a project of this size and type of construction. Plywood and timber varies widely in price - my only advice is to pick the best quality you can find. It may not be the cheapest, but will pay dividends during the life of the layout.

Above The undersides of the first two, station, baseboards are similar in that they provide low-level roadways front and back and a high-level trackbed in the centre. This view of a section of one of these baseboards shows more clearly than a thousand words of description the basic construction method employed. Viewed in conjunction with the other shots I think it will clarify the arrangement of supports, cross members and decking. The rear of the baseboard is to the left, the front to the right.

Above right The underside of the second baseboard showing a cross member and, on the right, the supporting piece of ply joined to the longitudinal section where the base rises from road level by the underbridge to track level at the joint with the third baseboard, the first of the two accommodating the fiddle yard. To the left can just be seen a vertical plywood support for the raised trackbed area.

Right A closer view of the same joint, but turned on its side and emphasising the simple nature of the carpentry that suffices. Incidentally, the paint-splattered softwood at the bottom of the picture is not part of the baseboard but the supporting frame running along the length of the garage and referred to in the text! The hole in the side of the baseboard (bottom right) is for the bolt and wing-nut that holds the legs in place when they are folded up; a similar arrangement holds the other side of the folding leg in place. Two such bolts are not absolutely necessary, but it does help to minimise the risk of accidental damage should the leg fall in transit - always a possibility with just a single bolt.

The basic structural parts of the baseboards are the sides, ends and cross pieces. Here is a close-up of a plywood cross member where it joins the side of the baseboard. Like other features of the baseboard, the cross members are supported with softwood bracing, and are pinned and glued to both the bracing and side members; the softwood strips, a minimum three-quarters of an inch square, are fixed similarly.

This is the standard arrangement for a corner joint. A three-quarter inch square softwood block is screwed and glued to the inside of the plywood longitudinal and end pieces. The softwood is cut to length to suit whatever other arrangements apply, for example where the trackbed or other areas use softwood as a support. Note that it is necessary to stagger the screws to avoid them hitting each other. Obvious, you might think - but alas I have on several occasions forgotten to do it!

Diagonal softwood cross members are cut around the corner joint softwood bracing and glued and pinned in place; there is little point in screwing into the end grain. This type of joint helps to ensure that the baseboard remains square and rigid. The softwood block just visible on the right is part of the arrangement for the leg pivots, while the ply block to the left is the strengthening for the pattern-makers' dowel.

Right The arrangement of a cross member supporting for the inside of the underbridge, seen from below. The angled plywood is the sidewall of the inside of the bridge, and is seen here from 'behind'. Once again the arrangement of the softwood supports is clearly seen, here also forming a solid base and fixing point for the roadway under the bridge.

Below Seen once again from below, the plywood forming the roadway under the bridge is now in place, supported on the cross members. The lines of screw heads show where the cross pieces forming the framework for the inside of the bridge are fixed as it crosses the baseboard.

Above Moving now to the upper side of the baseboard, here's a view of the underbridge from the front of the layout looking towards the fiddle yard, and showing almost the full extent of the baseboard. The sloping area is made from a separate off-cut of ply and the joint, although chamfered where it meets the flat section at the bottom, is also filled with plastic wood for a smooth transition and to minimise any problems when buildings, walls, pavements, etc, are added later. This area will become the site for a cinema and, being at the centre of the layout, with the eye drawn to this area because of the bridge and the shape of the raised area, it will be very important visually.

Left A closer view of the model bridge, the supporting plywood faces and the edge of the raised trackbed. Though a simple shot, it actually shows a few different aspects worthy of mention. The joint in the foreground shows the point where two pieces of surface board come together. The board providing the roadway under the railway goes across the full width of the baseboard, whereas the narrow, low-level surfaces along the front and rear of the baseboard are made from strips cut to suit their location and fixing.

The vertical trackbed support will form the basis for a retaining wall. Note that the corner joint at the bridge is pinned *and* glued; the pins are punched so that the heads do not protrude and cause problems later when the masonry detail is added.

Although perhaps not too clear from this picture, the vertical face curves slightly an inch or so from the bridge corner. This curvature is achieved by cutting a groove on the inside of the ply where the bend is to be made. This cut needs to go from top to bottom of the ply to be bent and, for a shallow bend such as this, a single cut, two-thirds of the thickness of the ply, is sufficient. A saw cut or even scoring a 'V' with a heavy-duty sharp knife will achieve the necessary groove.

A couple of prototype inspirations for this part of the model. The first photograph shows a sloping roadway and retaining walls, while the second provides a railside view of a plate girder bridge showing the masonry abutments. *Author/D. Hampson*

Two views of the front of the layout. The one above is looking into the fiddle yard. The track position is still being worked out, and the cardboard box represents a possible position for a signal box. In the lower view, looking towards the station, the site for the coal drops can be seen on the right.

Above The underbridge baseboard from the rear of the layout.

Below Turning now to the fiddle yard end of the baseboard, here is a view of the first of the two fiddle yard boards, that next to the open station area, looking from above the location of the sector plate out through the screen on to the layout proper. This picture shows the method of supporting the screen, the arc cut to allow the sector plate to swing, and the arrangement of the bracing for these open baseboards. As can be seen, there is one transverse cross member close to the interface with the sector plate, while the main bracing for this and its fellow fiddle yard board is a cross of softwood, which incorporates a cut out to accommodate and support the cross member. This all helps with the rigidity of the baseboard.

Left A closer view of a corner of the fiddle yard enclosure. The background of the photo is the vertical screen which goes two-thirds of the way across the baseboard to screen the entrance to the yard. As can be seen, this is fixed to a softwood strip which is in turn screwed and glued to the surface of the baseboard. A vertical strip provides a similar joint for the screen which runs along the front of the fiddle yard. The two vertical grooves in the right-hand screen enable the plywood to be bent to follow the shape required.

The baseboard surface has been cut in an arc to match the swing of the sector plate. Note that while the furthermost ply cross member is flush up to and supporting the fixed baseboard surface, the nearer softwood cross member is fixed lower to accommodate the deck of the sector plate which is made of thicker material to obviating the need for any bracing, which would complicate matters. The fiddle yard baseboards, being open-topped, require a slightly different approach to their construction.

Below left An overhead view of the arc cut in the surface of the fiddle yard baseboard to allow the sector plate to swing; the end of the sector plate must be similarly curved to match. So that the passage to and from the sector plate is as smooth as possible and alignment problems are minimised, the gap between the sector plate and the adjacent surface needs to be the minimum to allow the plate to swing. The pencil lines on the sector plate show where the fiddle yard roads will be located in order to match up to any of the three roads on to the station baseboards. I find that care in planning and marking this out is well rewarded.

Indeed, once the track positions are roughly marked on the baseboard, I would start the final tracklaying from this point. This transition from sector plate to railway proper is one of the key areas, if not *the* key area, in the tracklaying; once it has been established in the best position, the rest of the track can be finally fixed in place. The fiddle yard roads at their ends are, by the way, curved to match the entrance and exit roads as the sector plate is aligned with the main tracks.

Above right The sector plate itself is made from melamine-covered three-quarter-inch board, the melamine providing a good bearing surface. This view of the plate, inverted to reveal the underside, shows how a Meccano part has been screwed to it, the centre being over a pre-drilled hole in the board. This locates on to the pin epoxied on to the cross piece shown to the right; Meccano rod is useful. The two square pieces on either side of the pin are bearing surfaces taking some of the weight of the sector plate, and are very important in preventing the sector plate from tipping and keeping it level. The melamine coating provides an excellent hard-wearing and slippy bearing surface. It can be bought in iron-on strips of various widths for use where melamine-covered board is not used.

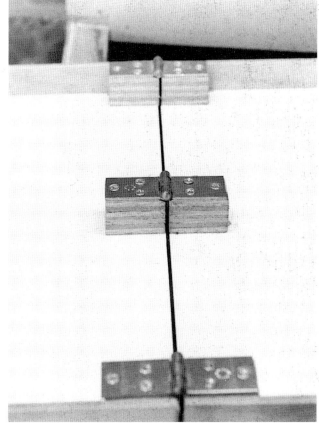

Right Because the sector plate is quite long and needs to cross two baseboards (thus obviating the necessity for 6-foot baseboards if trains of reasonable length are to be run), it is necessary to arrange for it to either split or fold. I chose to have it folding. The hinges need to be elevated above track level - no great height is necessary, just sufficient to clear the height of the track. The hinges are fixed to blocks of timber screwed and glued to the melamine; three are used because, although comparatively narrow, the sector plate is relatively heavy.

Left Moving to the station end of the layout, later we will see how the softwood timber frame for the platform sub-structure will be arranged; this is how the frame is arranged around the loco spur. The spur will have a buffer stop, a simple affair of a wooden plank fixed to the masonry. An alternative arrangement could, of course, be some type of rail-built structure. Alas, space is very limited, so we are looking for a suitable alternative that is both a possibility on the real railway and saves space. There are occasions when some form of cut-down rail-built arrangement can be seen, and this would be a compromise between the fully fledged rail-built buffer stop and the wooden balk.

The plywood rising vertically behind the loco spur runs along the length of the raised trackbed area and will form the basis of the parapet wall behind the platform. It will be cut away for the bridge platework and could also be cut away in the area away from the platform to provide an alternative to a wall - perhaps a steel-tube-and-cast-iron-post railing affair.

Right On the other side of the station, provision for the coal drops was necessary from the early stages of planning. Indeed, once having firmly decided on features such as this and the road underbridge, the rest of the planning, particularly for baseboard construction, must follow from the decision. However, even though the baseboards have been constructed, our joinery work is not finished. Once the exact location of the track over the coal drops is finalised and marked on the baseboard surface, the deck will be removed to allow the coal drop area to be constructed open as with the prototype. The sketch showing this feature (see pages 36-37) derives from the original LNWR proposals for this facility and will require the construction of brick piers and supporting steel girders spanning our newly created void, and timber balks, track and steel plates will be provided. Full details will be shown in the next volume of how this facility was constructed.

4.
LAYING THE TRACK

I will open this section with a confession - I have an aversion to building trackwork, any method, any scale. It is something that I try to avoid at all costs. This is not because I can't do it - far from it, I will do it if I have to, but I am a slow worker, I find trackbuilding tedious, and if I had to rely on building my own trackwork to produce a layout it would take forever or consist of one siding!

Each to his own, however. There are plenty who do enjoy trackbuilding and, indeed, there have been some very learned articles on the subject, together the practices of the railway companies in their tracklaying.

If you want to take trackwork seriously, the comments I made at the beginning of the book about following the practices of individual companies are equally valid. Individual companies did have their own practices in certain areas, for example, the practice of interlacing sleeperwork on points on the North Eastern or the different types of rail chair preferred by the different companies, usually related to the number of bolts used to fix them to the sleepers, and how they were arranged.

In the area of trackwork the finescale boys have been responsible for educating us all in the correct practices, and highlighting how track can be an important means of distinguishing for which company it has been laid. I do not pretend to be especially knowledgeable on matters of trackwork, and I would draw readers' attention to more detailed articles if they are particularly interested in these areas.

I know that my desire to get it *looking* right should also extend to trackwork, and I believe it does, but perhaps not to the extent that it should. In taking the holistic approach to the model, I console myself with the view that the trackwork is probably, providing it looks OK overall, the least important detail when it comes down, for example, to the type of chair used - it is the least noticeable feature and perhaps the area about which generally there is so far the least knowledge. OK, so it's a cop-out because of my aversion! Let's see what is available and what we can do to get the best.

Track choice

Starting at the bottom, 2 mm scale has its own track and wheel standards, but there is, at the time of writing, no ready-to-lay trackwork or pointwork available, although I believe it is possible to commission the construction of the latter. So far as I am aware components are available from the N and 2 mm scale societies who offer guides to its construction. Clearly the earlier comments about details of chairs and the like are hardly relevant in this scale, although having said that I've no doubt somewhere out there. . .

N gauge has a variety of track systems available from major manufacturers of products for that gauge. They are, with a few exceptions, generally capable of being used together, although between manufacturers radii vary amongst the ready-curved track and pointwork. In the UK Peco manufactures two extensive ranges of N gauge track, a standard range and a range with a rail section giving a finer appearance. These track systems are widely available here and abroad, and they are, unless you are still with a trainset or building your own, the gauge standards.

Similarly, for 16.5 mm gauge Peco produces two ranges of track, one finer in appearance than the other, with extensive pointwork, ready to lay and a whole range of supplementary products such as switches and point motors. The range is widely available. Peco also produces a do-it-yourself range which enables either traditional bullhead or later flat bottom rail to be produced to whatever gauge you are working in. A third rail system ready formed with moulded insulators is also available to represent, say, the Southern Electric system.

Other manufacturers of nearer scale (at least so far as sleeper spacing is concerned), ready-to-use trackwork include SMP and C&L. Both produce point kits and the latter offers ready-assembled points. The point kits of SMP are based on the tried and tested method of soldering rail to copper clad sleepers. Chairs of your choice need to be purchased separately and added cosmetically. The C&L system relies upon plastic moulded chairs being glued (or plas-welded) to plastic sleepers, then the rail is threaded through the chairs. Crossing vees are provided ready assembled to a choice of angles and the whole assembly on completion is strong, neat and very realistic at representing bullhead pointwork. Plans or templates are available separately and with both ranges a wide variety of pointwork is possible, so if you are prepared to design your own there should be little difficulty in copying any prototype formation. The C&L system provides very accurate and comparatively easy to build trackwork.

A number of manufacturers and specialist gauge societies provide the most common variations in cast chairs and other ironmongery which can either be cosmetic or functional. The traditional plywood sleepers that are riveted and then have rail soldered to the rivets are also available.

Moving to 7 mm scale, there are ranges of trackwork for O gauge (32 mm gauge) from C&L, mirroring that available for the smaller scales and including glue-on plastic fishplates. Peco produces readily available plain track and a limited range of point-

Most of the manufacturers of points and trackwork, and the components and kits for building them, provide plans showing full size their completed pointwork. Indeed, in many instances they are the plans from which the points are made. Usually manufacturers will sell these plans to enable you to plan your layout accurately, full size, prior to purchasing track and pointwork. A PECO 'Y' point template is shown here sellotaped down to the trackbed and is being used to determine the best position for the point and the way in which the plain track will line up with it. Templates and plans such as these are an invaluable planning aid and can take a lot of the guesswork and risk out of planning the layout and purchasing or making the pointwork.

Nonetheless, however careful I have been in using these plans I have always found that there is a need for some minor adjustment and realignment when the pointwork itself is actually put in place. Here the position of the pointwork at the station throat and for the loco spur has been determined, the alignment checked and the pointwork been pinned in place. The paper plan is being used to assess the final resting place of the point which will give access to Bollings Yard.

work which is much nearer to scale for British layouts in its sleeper spacing than its 16.5 mm counterpart and, carefully laid and painted, can be made to look very acceptable indeed, as the accompanying photographs show.

Slaters and others produce a variety of very detailed rail chairs and parts. An accurate model of Midland pointwork and track is easily made from the Slaters components, and the track construction uses wooden sleepers - smell that creosote!

For the project layout I used Peco O gauge track. The reason for this was that it is readily available and is most commonly used. It looks good when carefully laid and stands up well to the rigours of use on a portable layout that is kept in a garage where, despite being under the roof of the house, temperatures are to say the least variable.

Laying the track

Once the baseboard is complete, the track plan can be laid out upon it using the actual track and pointwork. Don't be surprised if at this stage things are not exactly as you planned - but there shouldn't be too much difference, perhaps a little more space in some areas, a little less in others. Also, as you stand before your baseboards, boxes of track in hand, you may be struck with some alternatives and amendments to the track plan that you hadn't thought of before.

Here is positively your last chance for a final decision on the track. Lay it out, join it together, put some rolling-stock on it. Represent the buildings with boxes and bits and pieces and see what you think of the layout full size.

When you are satisfied that it is right and to your

I think I may have flogged almost to death the matter of checking visually and physically the development of the layout at all stages, but it is particularly important on small and compact layouts that this is done religiously. It is surprising how only a minor adjustment can perhaps give that fraction more clearance which will allow an extra wagon in a siding or a slightly wider platform. Here an old brake-van with sizeable footboards is being used to check the clearance of the platform surface before the track is finally fixed in position. If you are laying track beside a curved platform, an easy way to check clearance is to take one of the longest vehicles you are likely to use, hold a pencil to the

centre of its side with the pencil point touching the baseboard surface, and push the vehicle round. The process can be repeated for the other side of the vehicle if there is a platform or wall on both sides of the track. The resultant pencil lines on the baseboard will tell you the nearest points to the track that any structure can be placed - remember, however, to allow for the overhang of the platform surface edging. It is also worth pointing out that outside cylinders on locomotives and long overhanging footplates can also cause problems. The cylinders of GWR locomotives, for example, tended to protrude further and required a wider loading gauge than other pre-nationalisation companies.

liking, that you have allowed for all reasonable train movements and left enough clearances everywhere, you can begin tracklaying in earnest.

Below While it is quite sufficient to pin plain, straight track to the baseboard every five or so sleepers, this is quite insufficient when it comes to securing the track across a baseboard joint. There are a number of ways of ensuring that when baseboards are split and re-assembled, track aligns properly and running is not impaired. One very satisfactory arrangement, when using track with a lighter section rail or in the smaller scales, is the simple expedient of soldering rail to screws or copper-clad material pinned and glued to the trackbed. This also has the advantage of being adjustable, should the need ever arise, by melting the soldered joint and realigning the rail. Note the removable-pin hinge joint referred to in the text.

Right The method of laying track across the baseboard joints of the project layout is shown here. The rail is laid across the joint, and the adjacent sleepers are moved to a position immediately beside the baseboard gap and pinned both centrally and at both ends. This method works quite well with O gauge track as used on this project. There is also a prototype justification for the positioning of sleepers closer together on either side of a

You will need to drill out every fourth sleeper or so to take a track pin, and to drill out the pilot holes in the pointwork sleepers to pin it down. Don't worry

baseboard joint, as the practice on the real thing was for closer sleeper spacing on either side of a rail joint.

Below right Only when the track is firmly fixed in place across the joint should the rails be cut. Cutting rails for any purpose is eased by the use of a carborundum slitting disc attached to a mini-drill. A mini-drill powered from a 12-volt supply (usually the power controller) is an indispensable and versatile tool for the railway modeller, enabling a number of cutting, grinding and polishing functions to be undertaken with the aid of a mandrel and suitable cutting tools in addition to the essential function of drilling. The powering of such a tool from a 12-volt supply enables it to be attached easily via mini crocodile clips to the running rails for power - particularly useful in dealing with emergency repairs. As with all tools, particularly power tools, even small lower-power ones, treat them with respect when in use to avoid accidents. Goggles might also be a wise precaution; the slitting discs shown here break quite easily and bits can spin off with remarkable speed and force - their edges are sharp, take it from me!

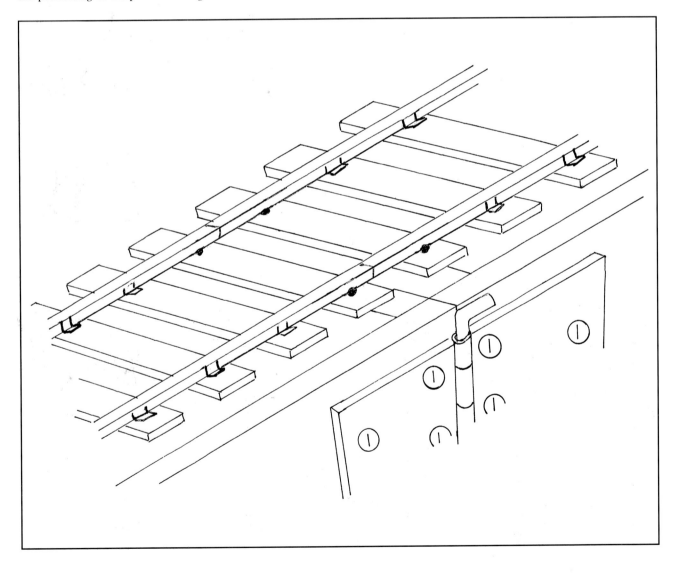

The proprietary track systems such as the Peco one used on Platt Lane have fishplates designed for that rail. These have the dual purpose of aligning rails across joints and conducting electricity from one piece of track across the joint to the next. Clearly the tighter the push-on fit of the fishplate over the rail, the better the electrical continuity; the length of the plate also helps this process. On occasion, however, it is not easy to allow the use of a fishplate at its full length without butchery to the adjacent sleepers, or even worse having them

spaced too far apart. One option is to carefully reduce the length of the fishplate with the aid of a carborundum slitting disc, as shown in use for cutting rail joints. Such an arrangement is shown here and obviates the necessity of cutting rather expensive ready-to-lay pointwork. Peco also provides nylon fishplates which can be used where an insulated rail joint is required and a metal fishplate could not be used. Use of nylon fishplates also helps to ensure good alignment across rail joints.

about electrical gaps or wiring at this stage - concentrate on getting the track pinned down to give a smooth transition into the pointwork and from straight to curve. Leave the pins slightly proud at this stage so that if you should need to move it you can get them out without damaging the sleepers. Try and get the trackwork to fit the pointwork which helps with the flow of line, and avoid over-tight curves when straightening, say, parallel sidings or loops as they come from the pointwork.

On the project layout I ensured that I had the right line for platforms and coal drops and fixed the pointwork in place, avoiding the baseboard joints, readjusting the track plan where necessary to get perfect alignment.

As mentioned earlier, where there is a traverser or sector plate I like to get the alignment of tracks between the fiddle yard and the layout proper as perfect as possible, and tend to spend a good deal of time at that point to ensure I get the best possible transition. It is necessary to fan out the sidings on a sector plate to ensure that they line up when the plate is swung to ensure alignment with the exit/entrance road. Be careful, however, not to put too tight a curve on the fiddle yard, and be prepared to be patient and to juggle track to get the best approach.

While I would normally put a pin in the sleeper centre every few sleepers on plain track, where the track crosses a joint or at the sector plate transition use two pins, one at each end of the sleeper adjacent to the joint. Lay the track across the joint, pin either side, then cut the track only when you are absolutely sure it is right.

You can now put some test wires on to the rails and, albeit with some limitations in the absence of full wiring, test out the layout, looking out particularly for smooth transitions and passage over pointwork and that you have achieved the necessary clearances. Make any necessary adjustments.

When you are completely satisfied, you can give some thought to the wiring proper.

An overview of the main station baseboard showing the track and platform tacked in their final positions for the main approaches, platform roads and loco spur; the positioning of the sidings and coal roads is still being finalised. The plywood face at the rear of the raised trackbed area will provide the base for the parapet wall above the rearmost retaining wall which rises from the street behind the station. The bottom left-hand corner of the picture shows the final position of the point giving access to the coal roads and siding. The positioning of this point is crucial, not only in relation to the positioning of the sidings, and particularly the coal drops for which provision has already been made in the baseboard for a cut-away area, but also because of the road underbridge and the edge of the raised section.

This edge will require a safety fence of some kind at this point, and accordingly it is necessary that the passage of the longest vehicles to be used over this section or the overhang of loco footplates will not foul any such provision. There are any number of similar matters that will arise - who said planning a model railway was simple? If you could predict all of them in advance of working out the final details full size, you would deserve to be awarded a PhD in model railway design.

5.
WIRING THE LAYOUT

Keep it simple, keep it tidy

Wiring the model railway layout is another one of those chores that has to be completed before we can actually enjoy running the trains or adding the scenics. However, like the baseboard construction and tracklaying, care in this very necessary stage will be rewarded later, hopefully with the minimum of electrical problems.

Let me begin by saying that I take a rather simple view of wiring and avoid 'high tech' approaches like the plague. Believe me, if I could get away with two wires, I would! I have experienced the difficulties of electrical problems in complex arrays of wire, switches and other gubbins applied layouts by electrical wizards who never seem to be around when their system fails! Accordingly my approach is based on two golden rules - keep it simple and keep it tidy.

The approach to wiring a permanent or portable layout is essentially the same, except that with the latter a means of easily connecting and disconnecting circuits across baseboard joints, from board to board, is needed.

I apologise to those of an electrical bent about to read the following paragraphs, but my views on and approach to the subject are based on a very basic 'need to know' understanding of the world of the electrician, and the experience that simplicity means reliability, particularly on exhibition and portable layouts. I am, however, the first to acknowledge that electrical wizardry can provide the ultimate in flexibility and control, and provide animation to appropriate areas of a model. Wiring a layout can be as simple or as complicated as you choose to make it.

To determine what the basic wiring needs of the layout are, we need to look at the track plan and the likely usage of it by the trains. Basically, the electrical supply and the wiring through which it is supplied can

- allow trains to move
- control the speed and direction of trains
- change the points
- operate the signals, and
- provide lighting to buildings, lamps, etc, on the layout.

The first two can be considered together, as they provide for the control and movement of the trains by supplying an electrical current from the controller to the motor in the locomotive via, usually, the track itself.

The electrical supply needs to be switched in and out of the various sections of the layout to allow maximum flexibility of use, and in particular the use of more than one locomotive at any one time. Similarly, if double track is to have any connection between the 'up' and 'down' line via crossovers or access is to be gained to loops and sidings from one line across the other, this needs to be planned for to avoid short circuits and other problems.

Connecting the power

The first point to consider is where to feed the electrical supply to the track and how.

Generally speaking, the electrical feed is arranged at the toe of pointwork so that on changing the point, the feed can power trains on either road leading out of the point. This situation is complicated further depending on how the points themselves are wired - of which more later.

At its simplest, the basis of all the power feeds to

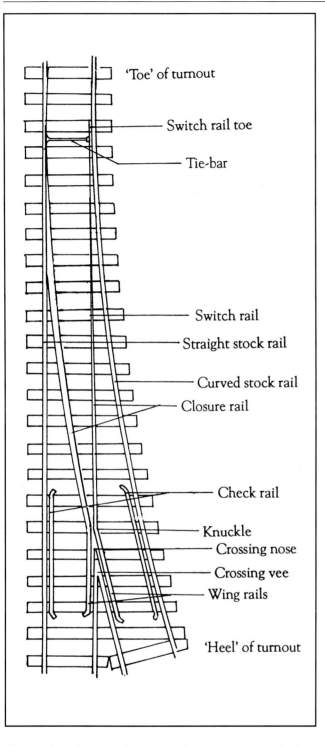

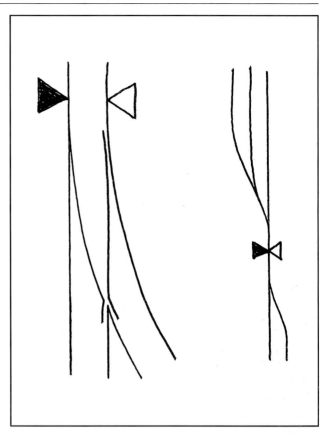

Left Anatomy of a simple turnout.

Above As a general rule of thumb, when planning the wiring of a layout arrange the power supply so that the feeds enable the current to flow from the 'toe' end of the turnout into the layout (left).
   One feed arranged as in the diagram on the right can service a number of routes, as used on the goods side of Platt Lane.

the track is that on the two-rail system, one rail takes the current to the locomotive and the other takes it back to its source - hence the term 'feed and return'. If, as has already been mentioned - and it is certainly the case with our project layout - we have more than one locomotive on the layout, we need to arrange certain sections of track to be switched on and off to enable trains to be stationary while others are run.

Such electrical isolation is easily accomplished by cutting a gap in one of the rails which is then bridged via a switch, thus allowing the current to be switched on and off to the section of track beyond the gap. There is a further possibility, touched on already, where the pointwork itself is wired so that it performs the function of switching the power to the outgoing track for which it is set; if a point is set for a siding, the current coming into it from the 'toe' end will flow into that siding, whereas when it is set for the other road it will flow to that one. This is the case with the Peco pointwork used on the project layout.

A point worth mentioning, but one that doesn't directly concern the project layout, is the particular problem of wiring reversing loops. I mention these for the sake of completeness and the fact that if I ever had the space, such an arrangement would be my ideal form of fiddle yard. The problem of electrical polarity means that a train cannot be taken round a reversing loop on one controller setting without it being stopped and the polarity being reversed by

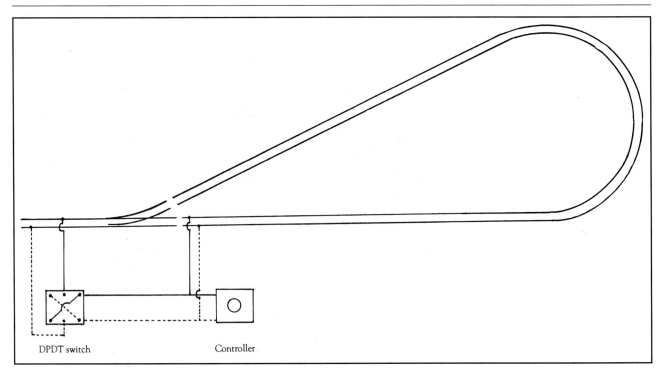

DPDT switch   Controller

Wiring for a reversing loop.

means of a double pole, double throw (DPDT) switch. I have enclosed a sketch of the arrangements for this.

The wire to use for model railway electrics should ideally be the multi-strand variety, as it is less prone to breakage than the single strand. The appropriate wire in a variety of colours is readily available from good model shops or radio parts stockists; if you can't find it locally, a good fall-back is the fine wire sold for auto electricians, readily available at motorists' accessory shops. It is a little thicker than we need, but nevertheless can be pressed into service.

Reference to the variety of colours available for such wire leads nicely into the area of colour coding. I use a red wire for all feeds and black for all returns. You can save a lot of wire by having a common return, ie a single-wire circuit linking all the returns on the circuits of one baseboard. Remember to link all the same side of the feed/return. The feed wires going to the tracks will need to be fed separately to each section via a switch, but can be linked between switch and controller providing the link takes place before the switch. Keeping to a particular colour of wire for a circuit helps ultimately in tracing any faults that may occur.

It is worth mentioning that the wiring is run under the baseboard to multi-pin sockets or switches and control panels. Tidiness comes in here - all the wiring should be harnessed together to form a 'loom' that can be fastened to the underside of the baseboards to avoid any damage from it hanging loose and being

caught. For this purpose I use plastic ties which are readily available from DIY, electrical and garden stores. I normally do not fasten them as tight as they will go, but just a notch off in case I need to either add another wire or, more likely, remove one which is to be replaced. To keep the loom tight under the baseboard, the ends of the ties should be pinned to the underside of the baseboard as close as possible to the wires. You may find that to aid this process it is necessary to cut holes in the baseboard cross members to keep the wiring up into the underframe and to enable the least obtrusive passage from track or point motor to socket.

Soldering

A digression on soldering would not go amiss at this point. Soldering is a basic technique that is impossible to escape, whether connecting wires to track or assembling metal locomotive and wagon kits. This latter aspect of soldering is dealt with in detail in the volume covering locomotive and rolling-stock construction. I will therefore concentrate in the next few lines on soldering techniques for wiring.

As mentioned earlier, the wiring of the layout is, like baseboards and trackwork, fundamental to the success of what comes later - museum-quality locomotives and trains will not realise their full potential if they don't run smoothly! The reliable supply of electricity to all the layout, as and when needed, is an essential requirement.

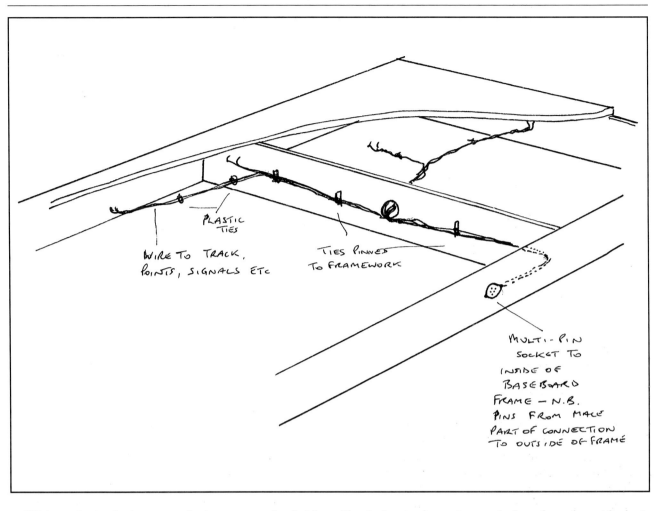

PLASTIC TIES

WIRE TO TRACK, POINTS, SIGNALS ETC

TIES PINNED TO FRAMEWORK

MULTI-PIN SOCKET TO INSIDE OF BASEBOARD FRAME — N.B. PINS FROM MALE PART OF CONNECTION TO OUTSIDE OF FRAME

Without doubt the best way of achieving good, reliable connections is to solder the wires, be it to trackwork, point motors, multi-pin connections or whatever.

Since you are not going to be able to get out of soldering, you may as well learn how to do it correctly to alleviate the fear. The two basic requirements are the cleanliness of the parts to be soldered, and having a soldering iron hot enough for the job. An iron with a small thin bit is best, allowing the joint to be made without damaging the surrounding area such as plastic sleepers or wire insulation. A soldering iron of 15-20 watts should be adequate for most electrical work.

Removing the plastic sleeving from the wire should reveal a clean core, ready to be 'tinned'. However, a rail or the contacts on a switch, etc, may need cleaning - a quick rub with a fibreglass burnishing tool or fine wet-and-dry paper will suffice.

To 'tin' the wire, apply the hot iron - hot enough to instantly melt the solder - and the solder itself (multi-core fluxed electrical solder) to the bared end of the wire to achieve a light but even coating of the wire with the solder. Allow it too cool - a few seconds only - then hold the tinned end in place either on the rail side or switch contact, threading it through any locat-

Sketch showing how wires can be looped together with plastic ties and fed through and along the baseboard frame to a central collecting point - the socket, or female half, of a multi-pin plug and socket. These pins and sockets, from 5 to about 30 pins, are readily availably from specialist electronics and radio parts suppliers.

Feeds to the track come from beneath the baseboard. Holes are drilled between sleepers immediately adjacent to the rails, and the wires are fed through from the underboard network and soldered neatly to the rail sides.

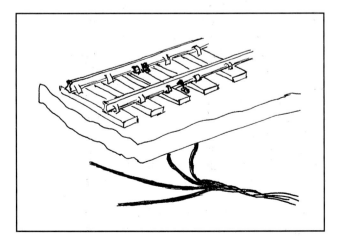

Above   The basic joint. The feed wire comes up through a pre-drilled hole as close as possible to the side of the rail, and is soldered to the railside. Only a small soldering iron is needed, enough to heat the rail and melt the solder - be careful not to leave the iron on the rail too long or it will damage plastic sleepers and chairs. I apply some flux to the railside where the joint is to be made, then apply solder, being careful not to get any on the top of the rail, and avoiding applying too much and producing a

'lumpy' joint. The bared end of the wire is then fed through the baseboard, held on the solder on the rail and the iron applied to the joint just long enough to soften the solder and make the joint.

Below   The equipment required for the basic wiring job: soldering iron, mild flux, electrical solder, wire-cutters, screwdriver and, of course, wire - lots of it, in different colours. Colour-coding the wire makes the tracing of electrical faults easier.

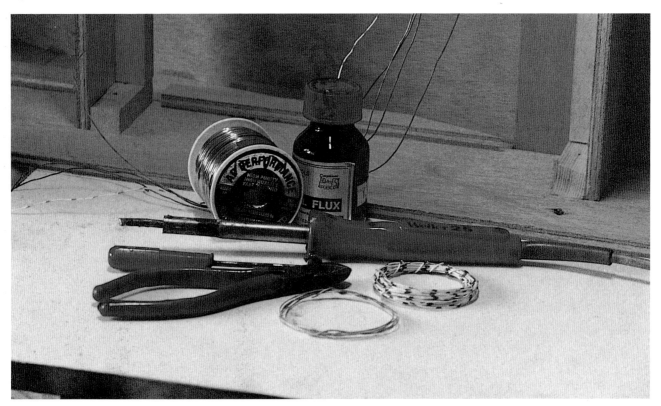

ing holes and applying the hot iron just long enough to melt the solder on to the rail or contact. Hold the wire in place after withdrawing the iron, allow it to cool, then give it a quick tug to make sure it is a good joint. If it isn't, repeat the process and perhaps add a mere touch of solder to the joint.

Control of Platt Lane

The plan shows the arrangements I have used for wiring Platt Lane. I have made provision for the main platforms, the approach and departure roads and the loco spur and siding to be operated from one controller. This, I appreciate, allows only one engine or train movement at any one time, but unless a regular number of additional operators were regularly available, I don't see practically that it would be very easy to either receive or dispatch a train simultaneously or carry out simultaneous engine movements while trains were departing or arriving, and operate the fiddle yard. You could wire in a second controller to any of the sections on either the up or down lines for this purpose quite easily.

I have, however, chosen to provide a second controller for the independent goods line with just one feed at 6. Additional switches and feeds could be provided at 7 with an insulating gap across both rails at X. This would enable shunting to take place at Bollings Yard while, say, a train left Platform 3 or coal

wagons were shunted in the coal drops; this again would require the use of two controllers. As we are providing two speed/direction controllers anyway, full flexibility for both the platforms and approaches and the goods roads could be provided if each controller could be switched into the other, but not the same circuits. The ideal option would be for controller number 1 to feed sections 1, 2, 3 and 4 and number 2 sections 2, 5, 6 and 7; section 2 would need to be arranged to allow controller 1 or 2 to be switched in as appropriate. My preference, and the one used here, is for one controller to be allocated to the station and one to the goods side - remember that the goods side is connected only to the other, passenger, side via the fiddle yard.

The arrangement for wiring the sector plate in the fiddle yard is shown earlier in chapter 2. The 'lead in' roads are wired to the entrance/exit roads 1 to 3, so the controller on the open or station baseboards controls access to and egress from the fiddle yard.

Some while back I mentioned the operation of points and signals. There are two alternatives, mechanical operation, which on the prototype is via angle cranks and rods to a lever frame (wire-in-tube on a model!), or electric operation via solenoids. The former is difficult to organise on a portable layout which has pointwork on more than one baseboard and needs to be operable both from front and back if you have one central control point.

For practicality I prefer the use of the good old sole-

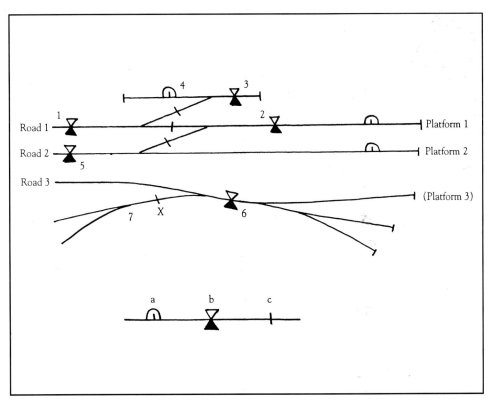

Wiring scheme for Platt Lane. The symbols used are:
a) a break in one rail only to create an isolated section at a siding or platform end. The gap is bridged by a simple on/off switch at the main control panel.
b) Feed of power from controller to track.
c) Break in both rails. The sections on either side of the break have separate, switched, feeds.

The construction of a simple control panel, which can be easily constructed from plywood and attached to the front or rear of the baseboard. The wiring from the panel connects to the layout via a multi-pin plug and socket, and is thus demountable.

The exploded view shows the pin-and-glue plywood assembly, the holes in the base for ventilation, the switches and 'mimic' diagram on the front face and the hole for the panel-mounted controller.

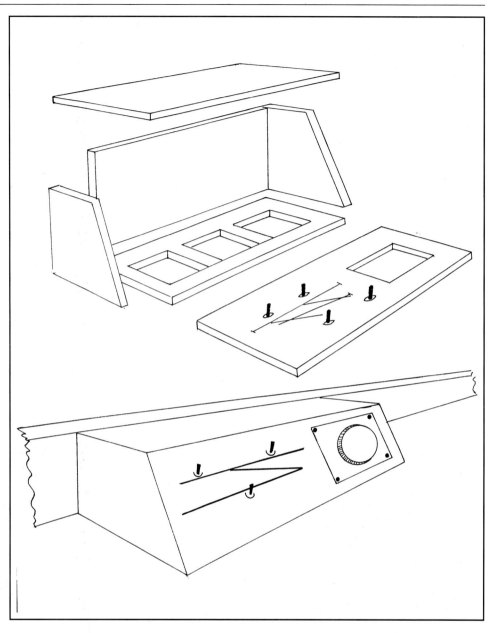

noid motor. Peco produce a motor for their pointwork which is very adaptable; it is capable of being clipped to the underside of the point, thus requiring the removal of an oblong of track bed to accommodate it below the baseboard. Alternatively, an adaptor is produced which enables the motor to be mounted above baseboard level, avoiding the need for the hole but requiring the motor to be hidden somehow. A third alternative, and the one used here and on my previous layouts, is to use the adaptor to mount the motor under the baseboard below the tie bar, linking motor to tie bar by a stiff wire which descends below the baseboard into the adaptor. This requires only a hole of some quarter of an inch in diameter to be drilled through the track bed exactly where the centre of the bar is located. Having located this exactly with the

aid of the point placed exactly in its final position, the point should be removed before the trackbed is attacked with a drill!

Wiring point motors could not be more simple. There are three wires to each solenoid - one to each coil and one to both coils. This last can be linked across to other solenoids, thus providing a common return. The other two wires are connected to the same switch, current passing down them to one or other coil and thus throwing the point. The switch used *must* be a passing contact switch, since electricity applied for more than a passing second to the coils will burn them out. Peco produces a passing contact switch which can be used on a mimic diagram control panel. Alternatively, if you like the idea of a bank of levers you could use those provided for the proprietary

An old H&M point motor, now only available on the second-hand market, used to operate a signal. It is fixed to the underside of the base-board by the simple expedient of bending the mounting bracket carefully at right angles. The wire from the signal to the angle crank can just be seen disappearing through a small hole drilled in the baseboard. The electric wire is multi-coloured and coded for ease of maintenance and repair and pulled together in a loom for safety as discussed in the text. The wires go to a multi-pin socket for connection to the next baseboard and the control panel.

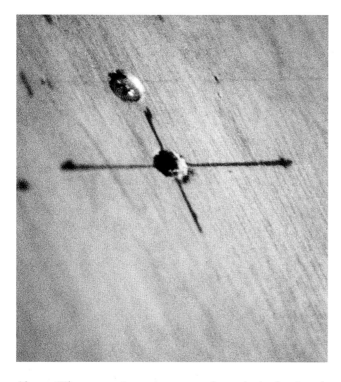

Above When mounting point motors beneath the baseboard, some form of link has to be made between motor and point. A hole must be drilled, and here we see how the position is marked using the axes of tie-bar and sleeper centre as reference points. The intersection of the two lines is the centre for drilling a 7/16-inch hole.

Below The Peco point motor mounted beneath a baseboard. Usually, and discounting auxiliary contacts for polarity, signal switching, etc, solenoid motors have four contacts for the basic operation. Those on one side can be linked together to form a common return. One wire is attached to each of the remaining contacts at each end of the solenoid, and the current is switched to these momentarily to change the point. It is essential that only passing-contact-type switches are used for switching point motors.

train set systems such as Hornby. I have a supply of old Hornby Dublo ones which I have used on small layouts. Each switch is numbered and the number of the switch is shown on a small track plan against the point or signal it operates - a bit like a real signal box.

Similar arrangements apply for operating signals; the accompanying photograph shows such an arrangement using an old H&M point motor, sadly no longer made. Note that the motor is placed end on to take advantage of the inbuilt crank. I believe that you really need a solenoid motor with a locking device for operating signals, which lets out the Peco unit for this purpose. SEEP and others produce such mechanisms.

Before leaving the subject of point and signal control it must be acknowledged that the use of solenoids for this purpose does produce a rather fierce movement of point blades and signal arms quite unlike that of the real thing. There are some quite expensive units available which utilise a small electric motor and worm-and-gear drive to a rod which pushes and pulls the tie bar. Similarly, there have been a number of articles in the model press over the years on how to get signal arms to 'bounce' on being pulled on and off. If this is an aspect that bothers you, I would refer you to those articles - so few modellers seem to bother with signals anyway, let alone working ones. A shame really, as signals

add character and ownership to a railway.

I also referred earlier to the lighting of buildings, etc. The use of fibre optics opens up lots of possibilities in this area and many splendidly detailed model building interiors have been illuminated and thus made visible by the use of nothing more than the good old 'grain of wheat' bulb. One particular area that interests me is the use of fibre optics to light semaphore signal lamps, and this is something that will be discussed in the next instalment.

Final thoughts

This particular volume has dealt with the least glamorous, least entertaining bits of building a model railway. For me, and, I guess, the majority, building baseboards, laying track and wiring a model railway is tedious both to do and, believe me, to write about. How do you make baseboard construction sound attractive and interesting?

In all seriousness, however, these are the fundamentals of any model railway and are the very basics upon

The next volume concentrates on the scenic development of Platt Lane. Here we see a mock-up of the station building - it will be quite some size!

which future operation and appearance depend. It is therefore worth going through the tedium to take care in planning out *your* model railway, building its foundation, laying the iron road and ensuring the means of propulsion is properly serviced. I cannot stress enough how important these aspects are - get the foundations right and half the battle is over. The next stages of the campaign are much more interesting.

I used the word 'your' a couple of sentences ago in the phrase '*your* model railway', for after all it is just that. You build what *you* want, run what *you* want. This series of books is designed simply to show you one approach based on looking at the real thing past or present, British or foreign, to provide an inspiration and a basis for your model rather than to build a model of a model. I hope that in discussing the evolution of the Platt Lane project you will be encouraged to look at the real thing for inspiration and perhaps discover, as I did, some gems on your doorstep.

The second volume in this series deals in detail with the construction of the buildings and the scenic details, both railway and non-railway, to complete the picture. Some have already been glimpsed in this volume. We will also look in detail at some of the things needed on model railways - but alas not on Platt Lane - such as trees and pasture land, to show how these can easily and effectively be modelled.

Finally, the last part of the series details almost blow by blow the building of the locomotives and rolling-stock used on the layout, the initial selection, the construction from brass, white metal or plastic kits, painting, lining, detailing, weathering and, last but not least, their usage.

In concluding this volume, I hope you have found it interesting, that it has perhaps provoked thoughts and encouraged you to have a go. While appreciating that it covers the least glamorous aspects of the hobby, I hope the balance will be redressed in the continuing story of our model railway project.

INDEX